Test Bank for

ORGANIC CHEMISTRY

WILLIAM H. BROWN

SHELTON BANK & JANET F. BANK

State University of New York, Albany

Saunders Golden Sunburst Series

SAUNDERS COLLEGE PUBLISHING

Harcourt Brace College Publishers

Fort Worth Philadelphia San Diego New York
Orlando Austin San Antonio Toronto
Montreal London Sydney Tokyo

Printed in the United States of America.

Bank/Bank: Test Bank to accompany Organic Chemistry by William H. Brown

ISBN 0-03-007578-5

56 017 98765432

PREFACE

This test bank accompanies the text, "Organic Chemistry" by William Brown. It contains 25 multiple-choice questions for each chapter of the text. Answers to the questions follow each chapter. This bank of questions is also available free from the publisher in computerized form for IBM and Macintosh computers.

IBM 3.5" disc	ISBN 0-03-007579-3
IBM 5.25" disk	ISBN 0-03-007589-0
Macintosh disk	ISBN 0-03-007592-0

Contents

Chapter 1 Covalent Bonding

1. The following species are isoelectronic. What is the electronic configuration that describes them?

   Mg^{2+} Na^{+} F^{-} O^{2-}

   a) $1s^2, 2s^2$ b) $1s^2, 2s^2, 2p^6$

   c) $1s^2, 2s^2, 2p^6, 3s^2$ d) $1s^2, 2s^2, 2p^6, 3s^2, 3p^6$

2. Which of the following species are isoelectronic?

Be^{2+}	Ne	S^{2-}	Mg^{2+}	N^{3-}
I	II	III	IV	V

   a) I, IV, V b) II, III, IV c) I, II, V d) II, IV, V

3. Arrange the following atoms in increasing order of electronegativity (least first).

   C O Li N K

   a) K, Li, C, N, O b) N, C, K, O, Li
   c) C, N, O, Li, K d) Li, K, C, N. O

4. Which of the following molecules contain both covalent and ionic bonds?

CH_3OH	Na_2CO_3	NH_4Cl	NaCl
I	II	III	IV

   a) I, III b) II, IV c) I, II, IV d) II, III

5. Arrange the following covalent bonds in increasing order of ionic character (least ionic first).

C-C	Na-O	C-N	O-H	C-O
I	II	III	IV	V

   a) III, I, IV, II, V b) V, III, I, II, IV
   c) I, III, V, IV, II d) I, III, II, IV, V

6. Which of the following Lewis structures is correct?

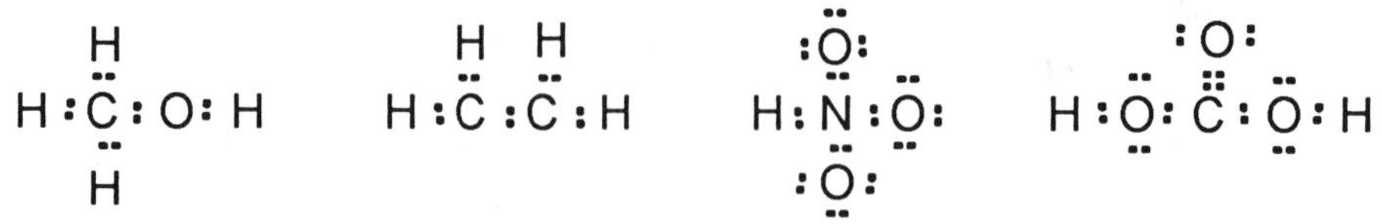

a) b) c) d)

7. Which of the following Lewis structures are correct?

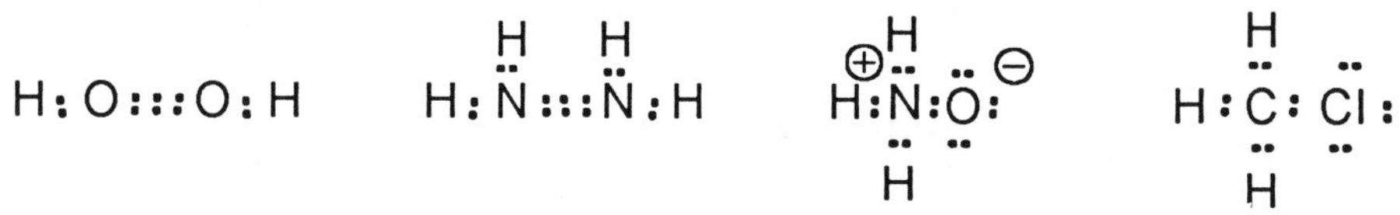

I II III IV

a) I, II b) II, IV c) III, IV d) I, III

8. Using the VSEPR model, predict which of the following species have bond angles of about 109°.

NH_3	CO_2	H_2O	$H_3O^{\oplus}$	O_3
I	II	III	IV	V

a) I, III, IV b) II, III, V c) I, IV d) III, IV, V

9. Using the VSEPR model, predict which of the labeled atoms have bond angles of 120° .

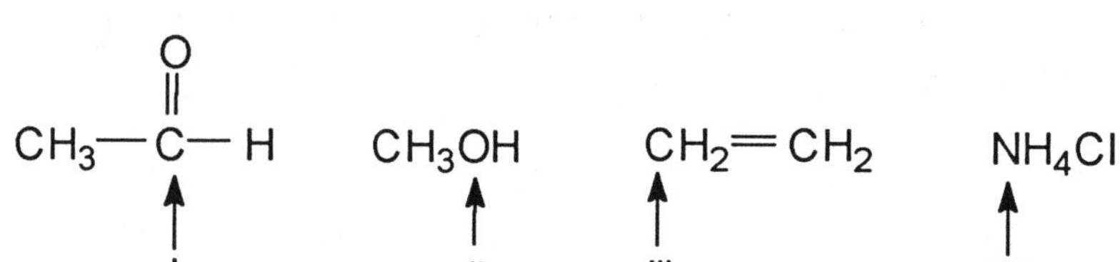

I II III IV

a) II, IV b) I, IV c) II, III d) I, III

10. According to VSEPR theory, what is your prediction for the arrangement of electrons for CH_3^- ?

a) linear b) tetrahedral c) bent d) trigonal

11. Which of the following functional groups have the correct Lewis structures?

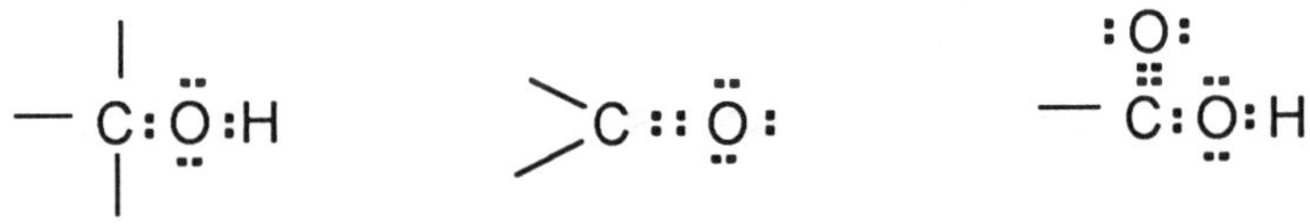

hydroxyl I carbonyl II carboxyl III

a) I, II b) II, III c) I, II, III d) I, III

12. What is the correct structure for the aldehyde which has the formula C_4H_8O ?

a) 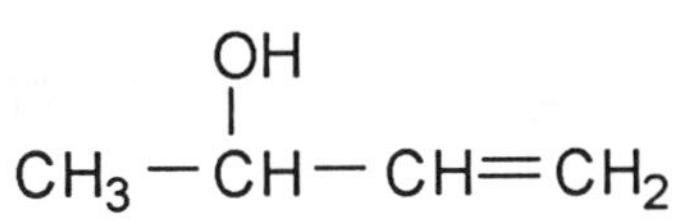

b) 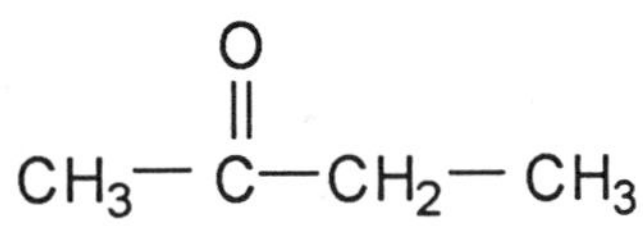

c) 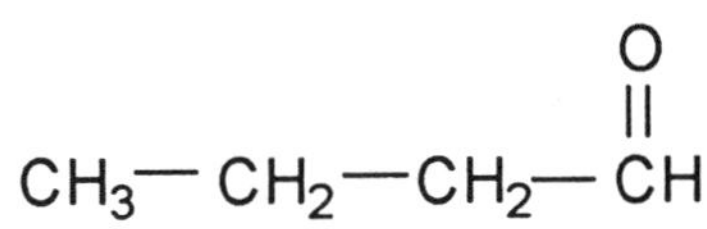

d) $CH_2=CH-CH_2-O-CH_3$

13. What is the correct Lewis structure for acetic acid ($C_2H_4O_2$) ?

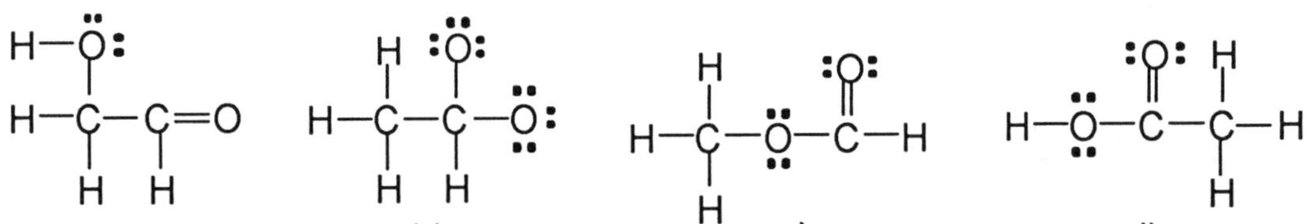

a) b) c) d)

14. Which functional groups are in the following molecule?

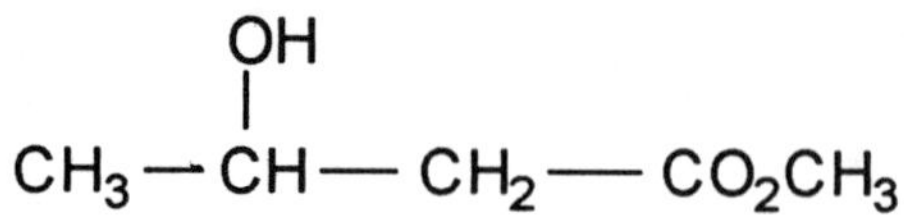

a) alcohol and ester b) alcohol and acid
c) ketone and ether d) ketone and aldehyde

15. Which of the following pairs of molecules are constitutional isomers ?

a) $CH_3CH_2OCH_3$ and CH_3CH_2CHO

b) $CH_2CH_2OCH_3$ and CH_3COCH_3

c) CH_3CH_2CHO and CH_3COCH_3

d) CH_3CH_2OH and CH_3CHO

16. Which of the following belong in the group of constitutional isomers for $C_3H_6O_2$?

I. $CH_3CO_2CH_3$ II. $CH_2{=}CHCH_2OH$

III. $CH_2{=}CHCO_2H$ IV. $HOCH_2CH_2CH_2OH$

V. CH_3OCH_2OH

a) II, III b) II, IV, V c) I, II, V d) I

17. Which of the following pairs of molecules are constitutional isomers?

a) 

b) 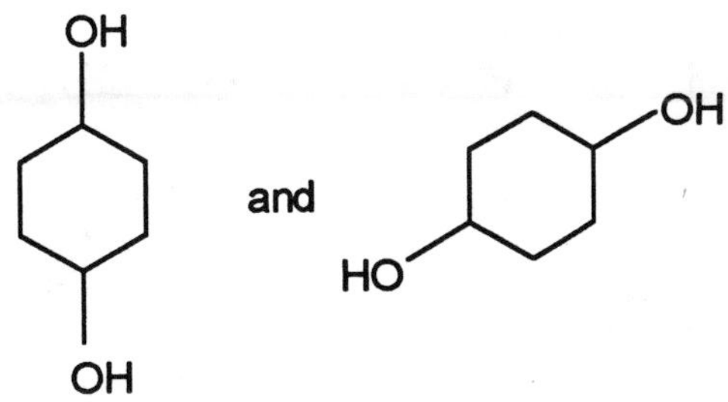

c) 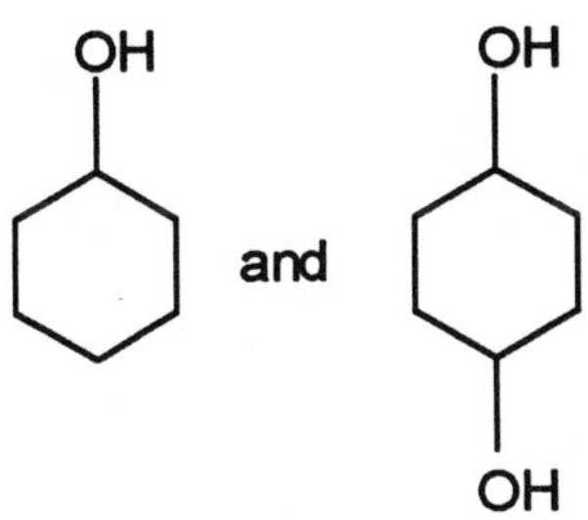

d) 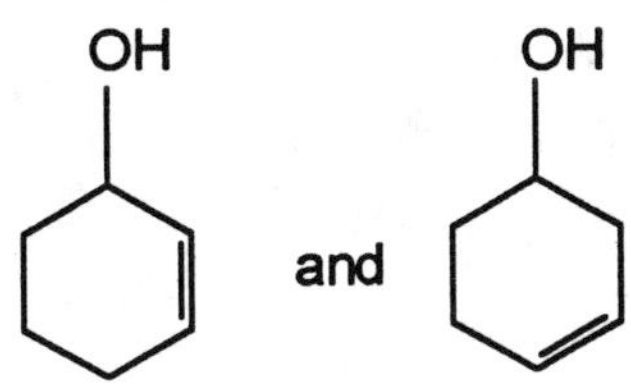

18. Which of the following are contributing resonance structures for the bicarbonate ion ?

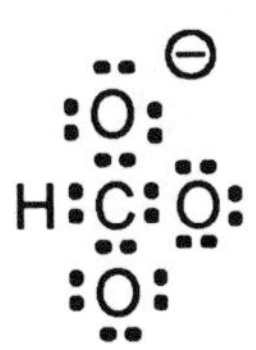

I

II

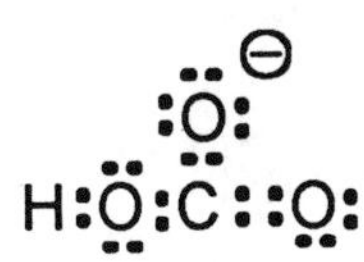

III

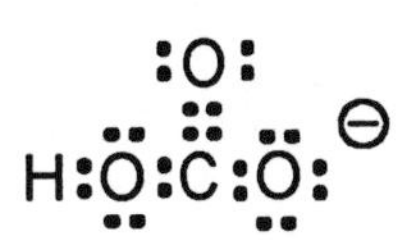

IV

a) I, III   b) II, III   c) I, IV   d) III, IV

19. In each set of contributing resonance structures, which structure makes the greater contribution?

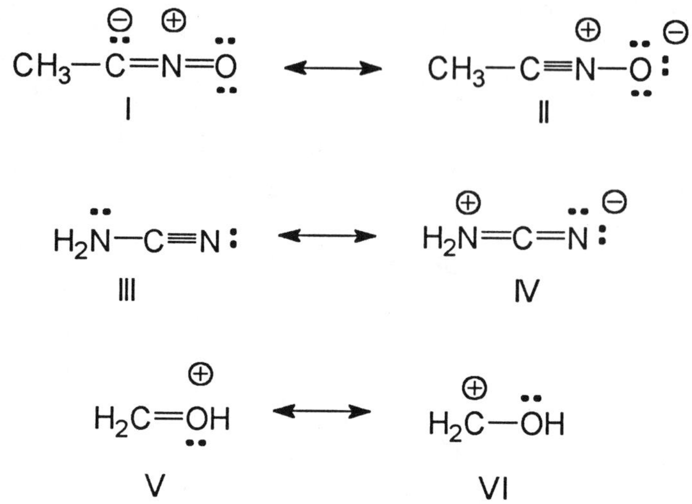

a) I, III, V b) II, IV, VI c) I, IV, V d) II, III, V

20. Which of the following are correct pairs of contributing resonance structures ?

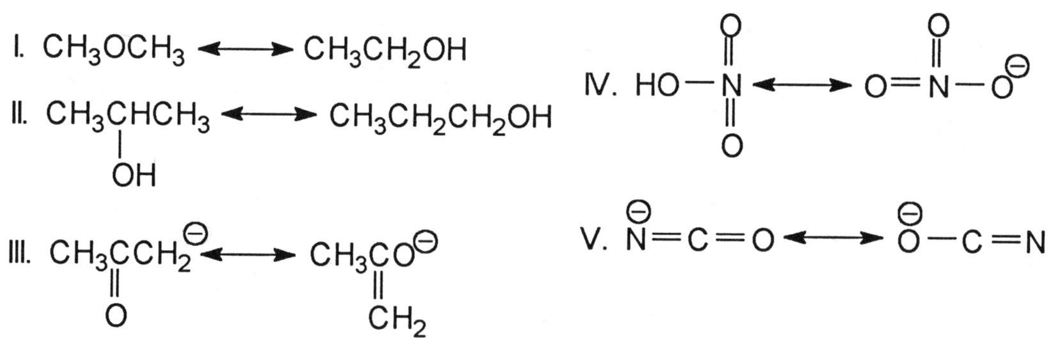

a) III, IV b) I, II c) IV, V d) III, V

21. Which of the following pairs are contributing resonance structures ?

I. $CH_3OCH_3 \longleftrightarrow CH_3CH_2OH$

II. $CH_2{=}O \longleftrightarrow \overset{\ominus}{C}H_2{\equiv}\overset{\oplus}{O}$

III, $CH_2{=}CH{-}\overset{\ominus}{C}H_2 \longleftrightarrow \overset{\ominus}{C}H_2{-}CH{=}CH_2$

IV. $\overset{\ominus}{N}{=}C{=}O \longleftrightarrow N{\equiv}C{-}O^{\ominus}$

a) II and IV b) I, II, III c) III and IV d) II, III, IV

22. What is the correct orbital hybridization for the following carbon atoms ?

$H_2C{=}O$	$H_2C{=}CH_2$	$H_3\overset{\oplus}{C}$	$HC{\equiv}N$	$O{=}C{=}O$
sp	sp^2	sp^3	sp	sp
I	II	III	IV	V

a) I, II, III b) II, III, IV c) II, IV, V d) I, IV, V

23. What are the correct orbital hybridizations for carbon in the following species ?

A. $CH_3^{\ominus}$	I. sp
B. CH_4	II. sp^2
C. $CH_3^{\oplus}$	III. sp^3

a) A and I, B and III b) B and I, C and II
c) A and III, C and II d) B and III, C and III

24. In which of the following molecules does carbon have both sigma and pi bonds?

CH_4	C_2H_2	H_2CO	CH_3NH_2	CO_2
I	II	III	IV	V

a) II, V b) I, IV c) II, III, V d) II, III

25. Which of the following represent bonding for a three carbon system?

I

II

III

IV

a) I, III b) I, IV c) II, III d) II,IV

Answers

1. b)
2. d)
3. a)
4. d)
5. c)
6. d)
7. c)
8. a)
9. d)
10. b)
11. d)
12. c)
13. d)
14. a)
15. c)
16. d)
17. d)
18. d)
19. d)
20. d)
21. c)
22. c)
23. c)
24. c)
25. b)

Chapter 2 Alkanes and Cycloalkanes

1. What is the correct IUPAC name for the following compound?

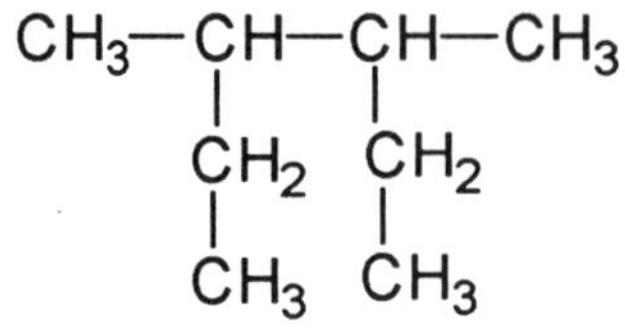

a) 2-ethyl-3-methylpentane
b) 3,4-dimethylhexane
c) 2,3-diethylpentane
d) 3-methyl-4-ethylpentane

2. What is the correct name for a constitutional isomer of C_4H_9Br ?
a) 2-methyl-2-bromopropane
b) 2-bromo-2-methylbutane
c) 1-bromo-2-methylpropane
d) 3-bromobutane

3. What is the correct IUPAC name for the following cycloalkane?

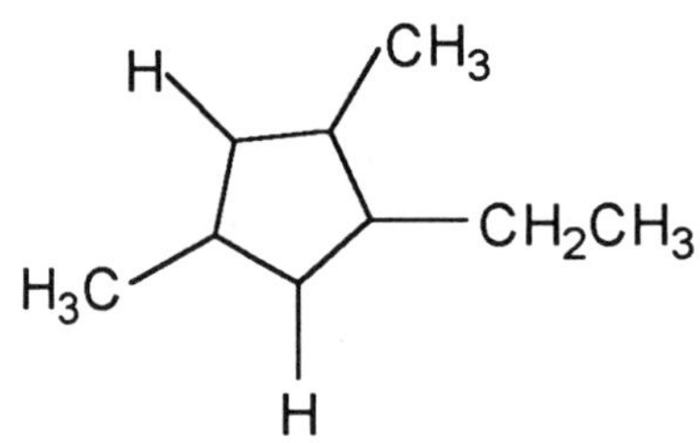

a) 2,4-dimethyl-1-ethylcyclopentane
b) 1,3-dimethyl-5-ethylcyclopentane
c) 1-ethyl-2,4-dimethylcyclopentane
d) 1-ethyl-3,5-dimethylcyclopentane

4, Which of the following matches are correct?

A. 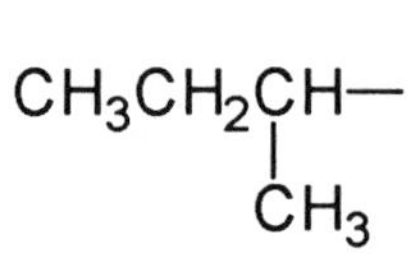

C. 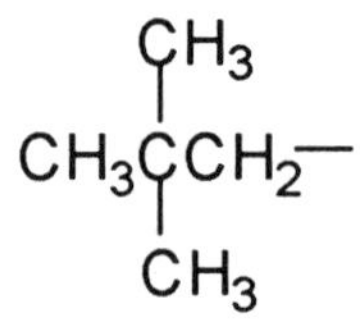

B. 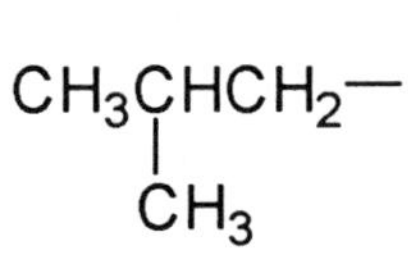

D. 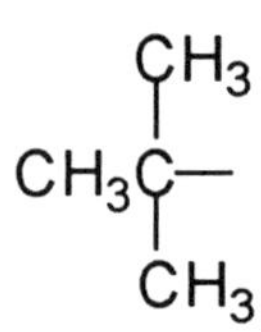

I. tert

II. iso

III. sec

IV. neo

a) A and III, B and II
b) B and III, D and I
c) C and I, D and IV
d) A and II, C and IV

5. Which isomer of pentane has the lowest boiling point?

$CH_3CH_2CH_2CH_2CH_3$ $(CH_3)_2CHCH_2CH_3$ $(CH_3)_2C(CH_3)_2$ 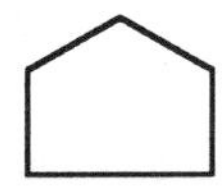

a) b) c) d)

6. Arrange the following hydrocarbons in the order of increasing boiling point (lowest first).

I. gasoline
II. lubricating oil
III. diesel fuel
IV. LPG
V. kerosine

a) V, I, IV, II, III b) IV, I, V, III, II c) I, V, IV, II, III d) IV, V, I, III, II

7. Place the following species in the order of increasing energy of attractive forces (lowest first).

I. CH_3-CH_3
II. H_2O-H_2O
III. $Na^+ Cl^-$
IV. Ar−Ar

a) IV, II, I, III b) III, II, IV, I c) I, III, II, IV d) IV, III, I, II

8. How many constitutional isomers are there for $C_5H_{11}Br$?

a) 3 b) 7 c) 8 d) 11

9. Which of the following are constitutional isomers of 4-isopropyloctane?

I. 3-ethyl-2,4,5-trimethyloctane
II. isobutylcyclohexane
III. 2,2-dimethyl-4-ethylheptane
IV. 4-ethyl-2-methyloctane

a) I and IV b) II and III c) I and II d) III and IV

10. How many isomers (constitutional and geometric) are there for dimethyl-cyclopentane?

a) 3 b) 5 c) 7 d) 9

Chapter 2 Alkanes and Cycloalkanes

11. Which of the following is a conformer of n-pentane?

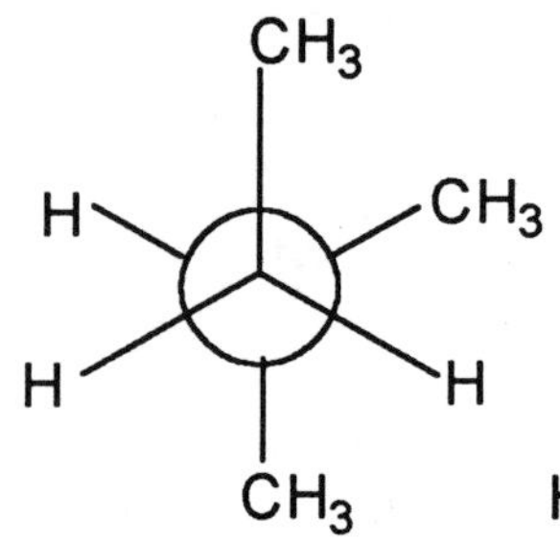

a)

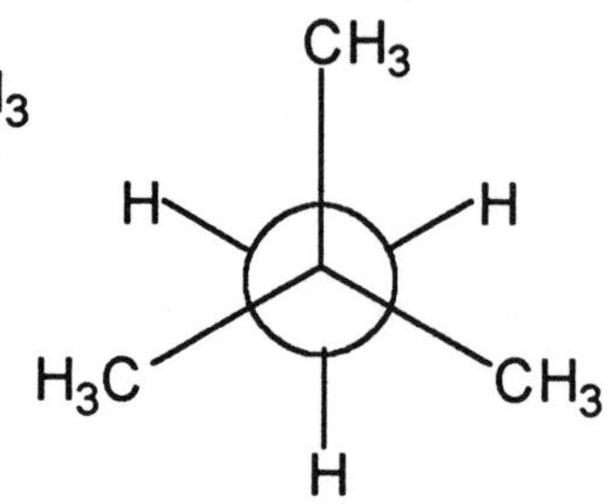

b)

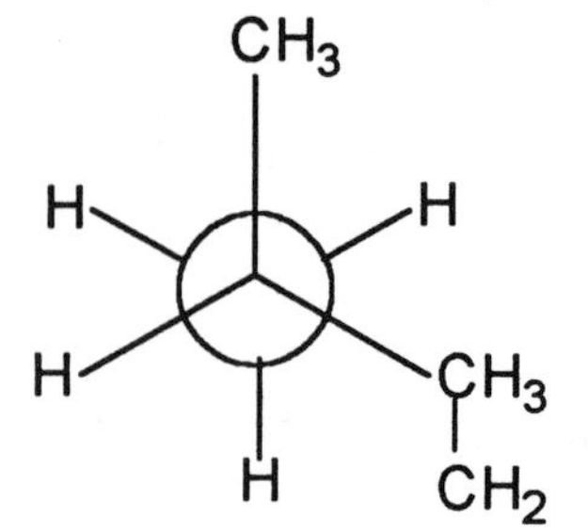

c)

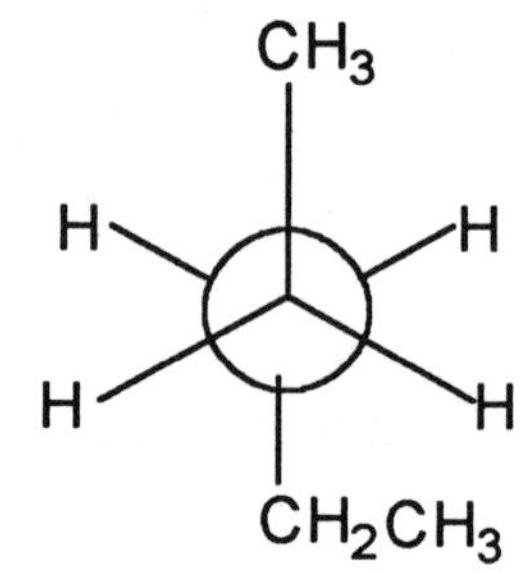

d)

12. What is the order of increasing stability (least first) for the following conformers of 2,3-dimethylbutane?

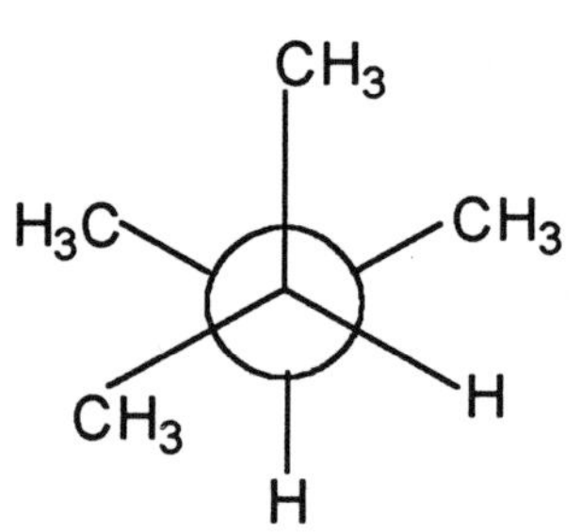

I

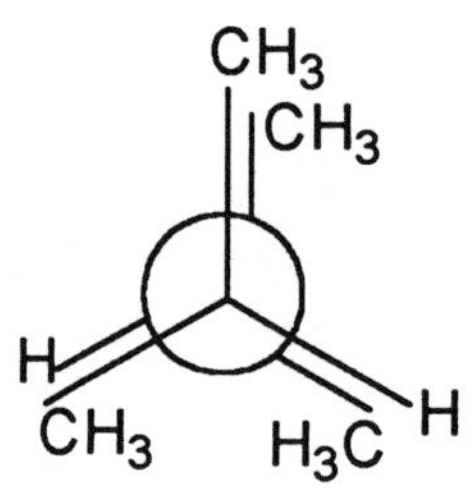

II

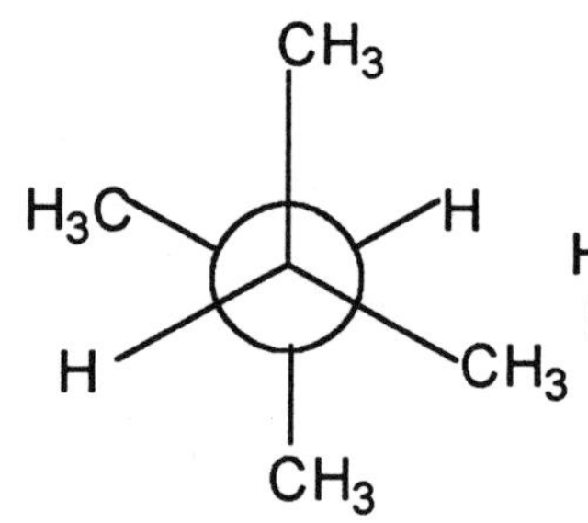

III

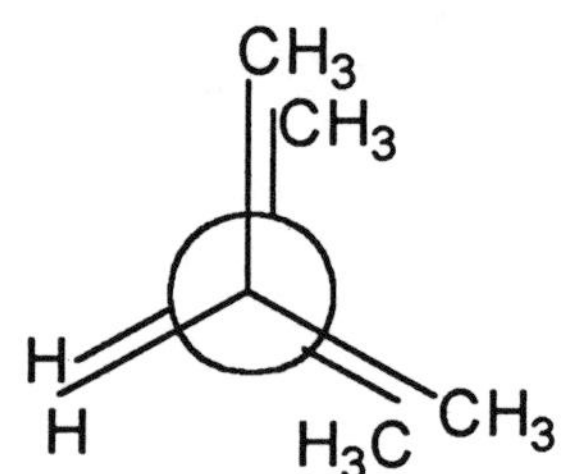

IV

a) I, II, IV, III b) IV, I, III, II c) II, IV, III, I d) IV, II, I, III

13. Which of the following conformers of isobutyl chloride is the most stable?

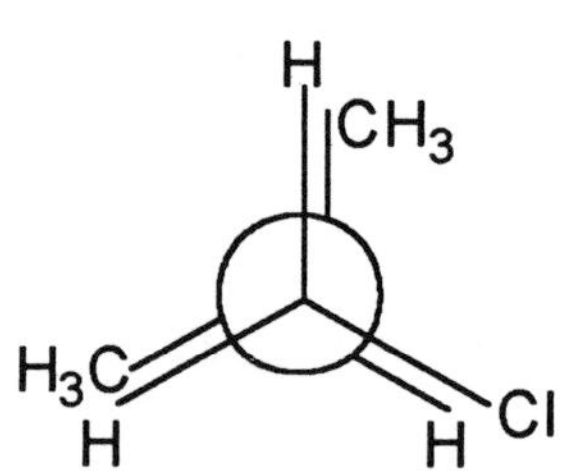

a)

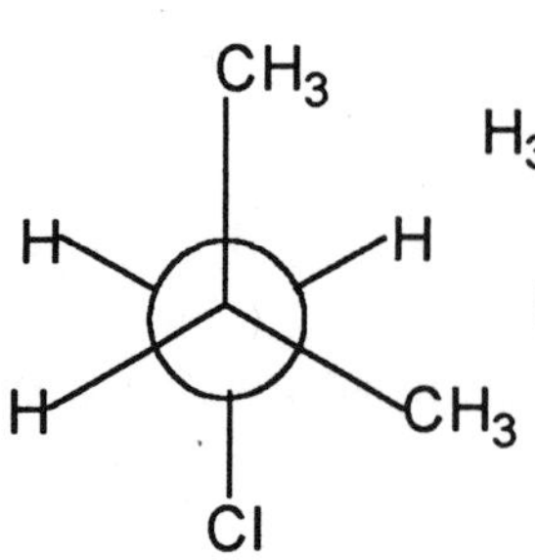

b)

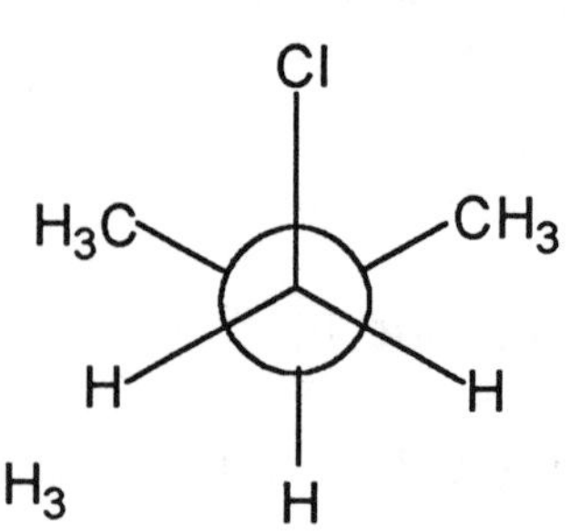

c)

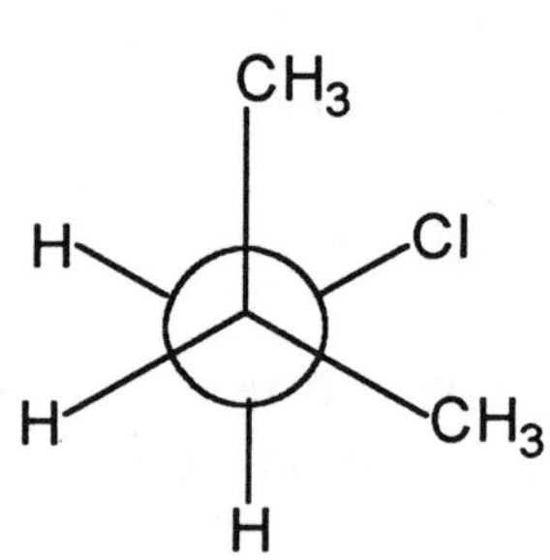

d)

14. Which of the following is the gauche form of propyl bromide?

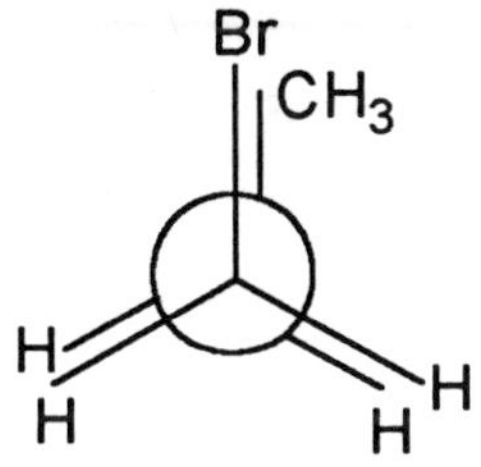

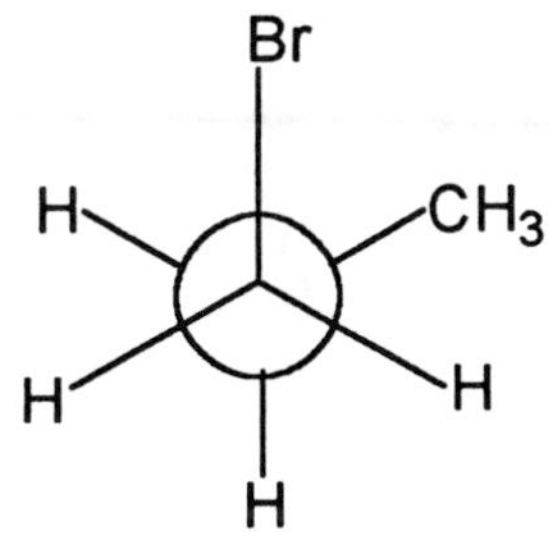

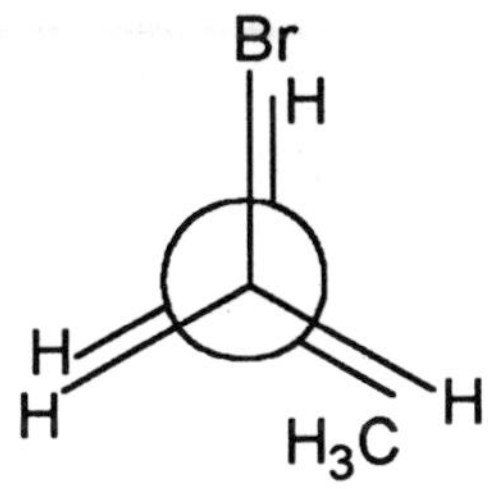

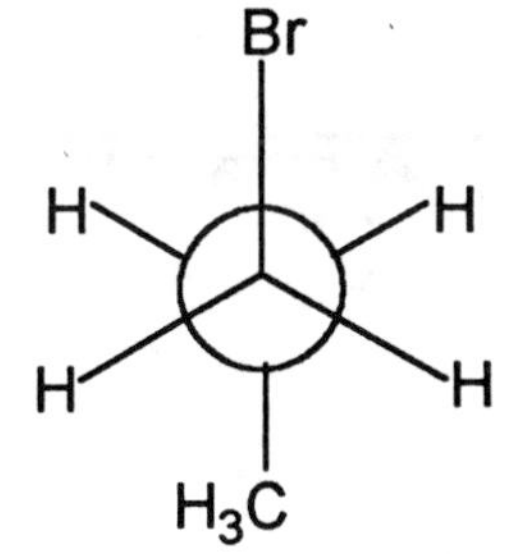

a) b) c) d)

15. Which structures are the cis form of 1,2-dimethylcyclohexane?

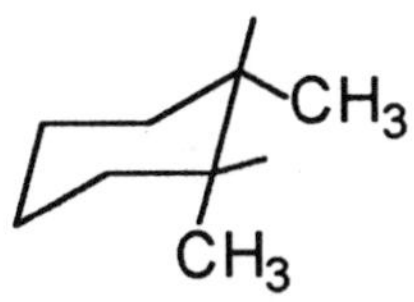

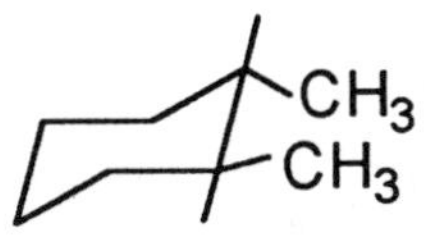

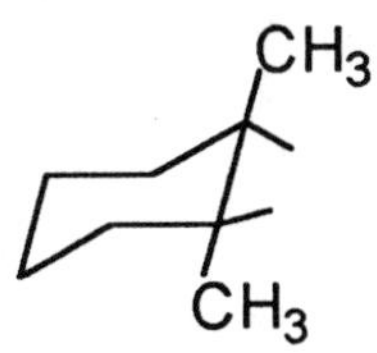

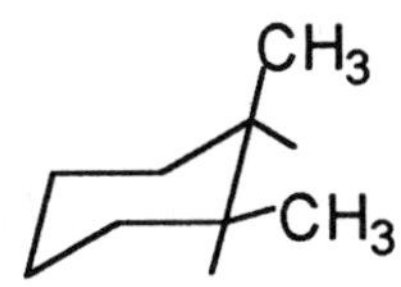

I II III IV

a) I and IV b) II and III c) III and IV d) I and II

16. In the most stable conformation of trans-1,4-dimethylcyclohexane, what positions do the methyl groups occupy?

a) axial, axial
b) equatorial, axial
c) equatorial, equatorial
d) axial, equatorial

17. Which of the following are more stable isomers?

I. cis or trans 1,4-dibromocyclohexane
II. cis or trans 1,3-dibromocyclohexane

a) I cis and II cis
b) I cis and II trans
c) I trans and II cis
d) I trans and II trans

18. For which of the following compounds is the cis form more stable than the trans form?

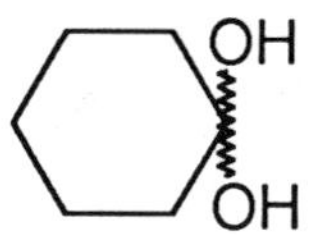
a)

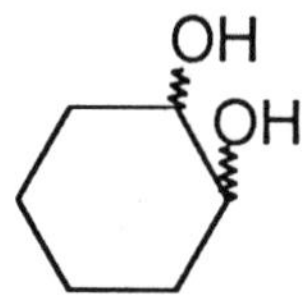
b)

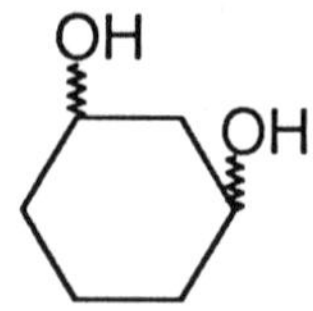
c)

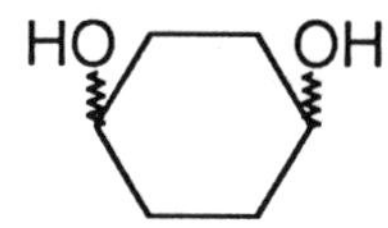
d)

19. Which of these diaxial forms has the highest energy?

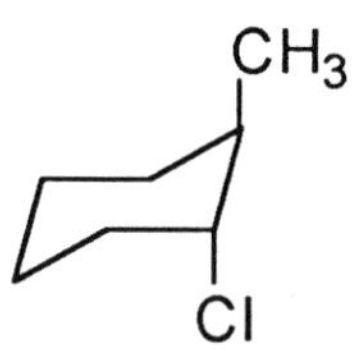
a)

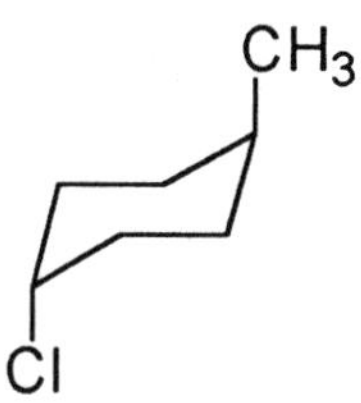
b)

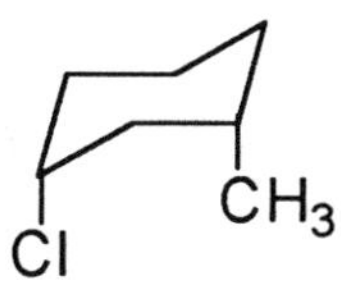
c)

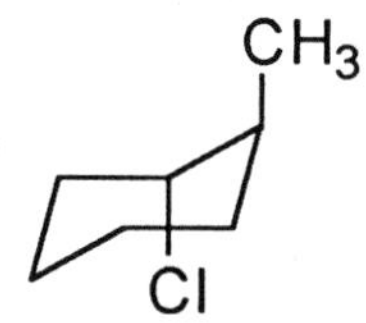
d)

20. Which constitutional isomer of hexane will give the greatest number of different monochlorinated products?

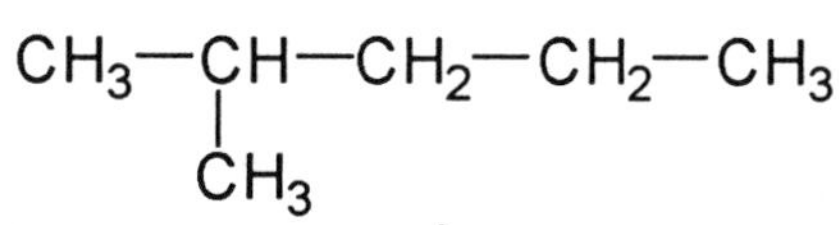
a)

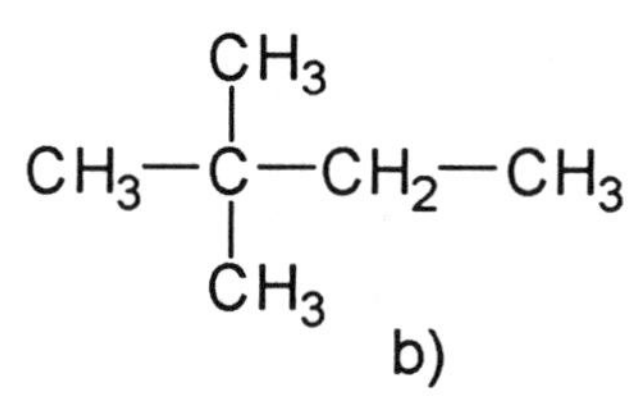
b)

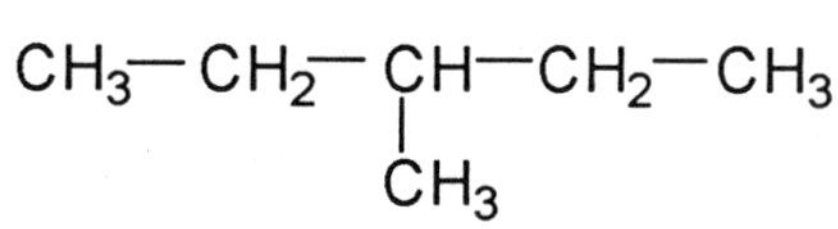
c)

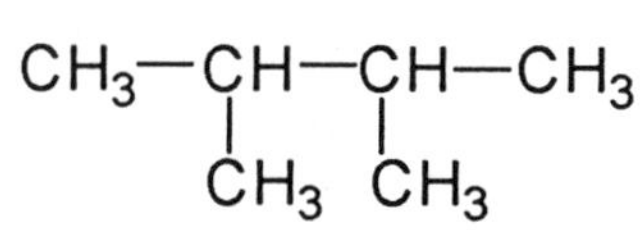
d)

21. How many monochlorinated products would result from the following reaction?

3-methylpentane + chlorine + light or heat ⟶

a) 3 b) 4 c) 5 d) 6

22. Which compound will give 4 different monochlorinated products?

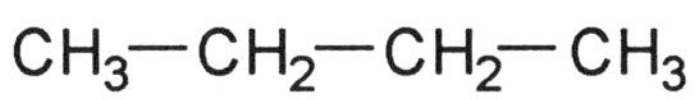

a)

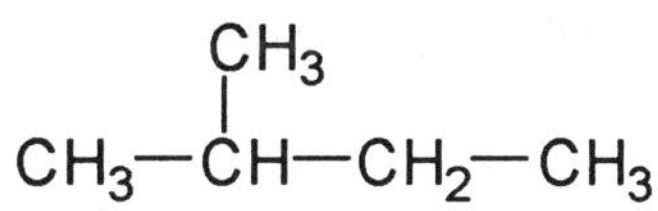

b)

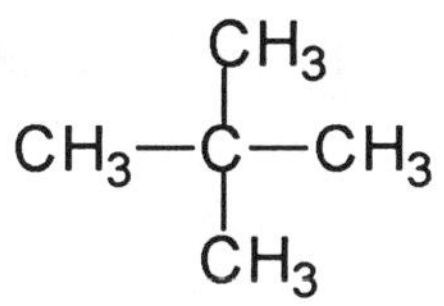

c)

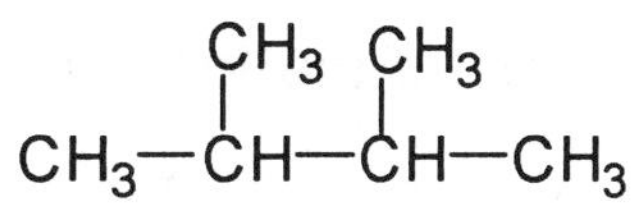

d)

23. Which constitutional isomer of C_5H_{12} gives a single monochlorinated product when reacted with chlorine and heat or light?

a) pentane b) isopentane c) neopentane d) cyclopentane

24. When 2-methylpentane is reacted with bromine and light or heat, which compound is formed with the greatest regioselectivity?

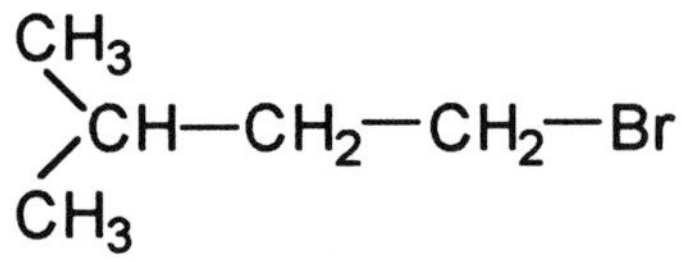

a)

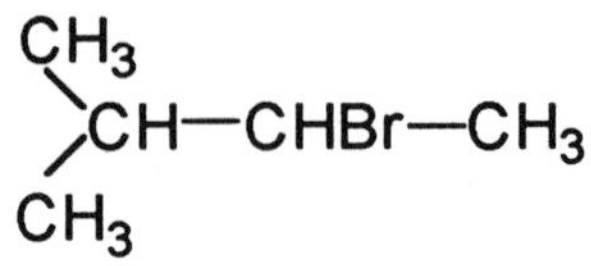

b)

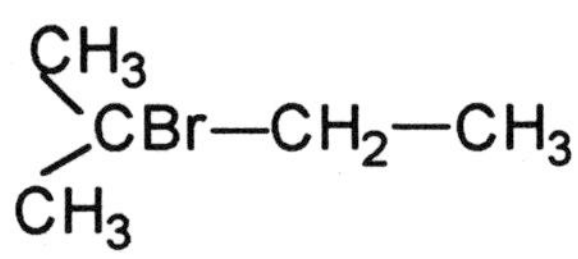

c)

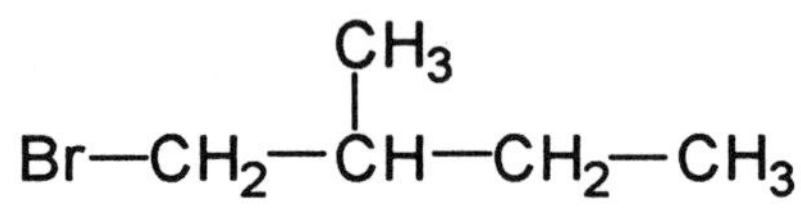

d)

25. What is the name of the following commercial process in the fuel industry?

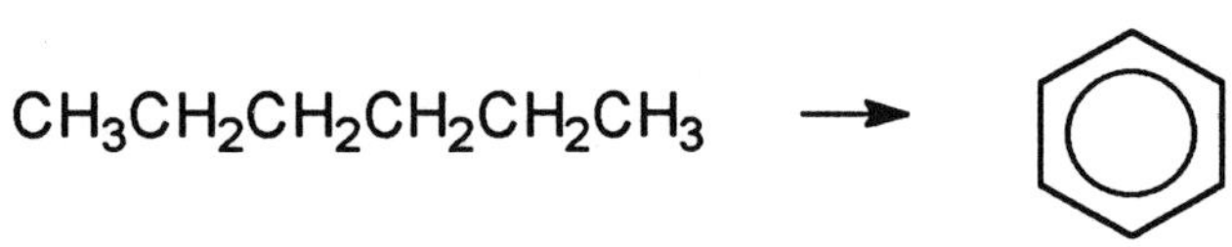

a) Fischer-Tropsch process

b) antiknock process

c) alkylation

d) catalytic reforming

Answers

1. b)
2. c)
3. c)
4. a)
5. c)
6. b)
7. a)
8. c)
9. d)
10. b)
11. d)
12. d)
13. b)
14. b)
15. a)
16. c)
17. c)
18. c)
19. c)
20. a)
21. b)
22. b)
23. c)
24. c)
25. d)

Chapter 3 Acids and Bases

1. Which of the following substances are bases according to the Lewis theory?

I	II	III	IV
$CH_3-CH=CH_2$	$CH_3-\underset{CH_3}{\overset{CH_3}{\overset{\vert}{\underset{\vert}{C}}}}-Cl$	BF_3	$CH_3-\overset{\oplus}{C}H-CH_3$

a) I, II b) III, IV c) II, III d) I, III

2. Which of the following substances are bases according to the Lewis theory?

I	II	III	IV
H_2O	$AlCl_3$	$CH_3\overset{\oplus}{C}HCH_3$	$\overset{\ominus}{F}$

a) I, II b) I,III c) III, IV d) I, IV

3. Identify the Bronsted-Lowry acids in the following reactions.

$$\underset{I}{CH_3OH} + \underset{II}{HCl} \rightleftharpoons \underset{III}{CH_3\overset{\oplus}{O}H_2} + \underset{IV}{\overset{\ominus}{Cl}}$$

$$\underset{V}{C_2H_5OH} + \underset{VI}{\overset{\oplus}{Na}\overset{\ominus}{H}} \rightleftharpoons \underset{VII}{C_2H_5\overset{\ominus}{O}\overset{\oplus}{Na}} + \underset{VIII}{H_2}$$

a) I, III, VI, VII b) II, VI c) I, IV, V, VIII d) II, III, V, VII

4. Identify the conjugate bases in the following reactions.

$$CH_3NH_2 + CH_3CO_2H \rightleftharpoons \underset{\text{I}}{CH_3\overset{\oplus}{N}H_3} + \underset{\text{II}}{CH_3C\overset{\ominus}{O}_2}$$

$$C_2H_5OH + \overset{\oplus}{Na}\overset{\ominus}{H} \rightleftharpoons \underset{\text{III}}{C_2H_5\overset{\ominus}{O}\overset{\oplus}{Na}} + \underset{\text{IV}}{H_2}$$

$$CH_3OH + HCl \rightleftharpoons \underset{\text{V}}{CH_3\overset{\oplus}{O}H_2} + \underset{\text{VI}}{\overset{\ominus}{Cl}}$$

a) II, III, VI b) I, IV, V c) I, III, V d) II, IV, VI

5. Identify the conjugate acids in the following reactions.

$$CH_3NH_2 + H_2O \rightleftharpoons \underset{\text{I}}{CH_3\overset{\oplus}{N}H_3} + \underset{\text{II}}{\overset{\ominus}{O}H}$$

$$CH_3OH + HCl \rightleftharpoons \underset{\text{III}}{CH_3\overset{\oplus}{O}H_2} + \underset{\text{IV}}{\overset{\ominus}{Cl}}$$

$$C_2H_5OH + \overset{\oplus}{Na}\overset{\ominus}{H} \rightleftharpoons \underset{\text{V}}{C_2H_5\overset{\ominus}{O}\overset{\oplus}{Na}} + \underset{\text{VI}}{H_2}$$

a) I, IV, VI b) I, III, VI c) II, IV, V d) I, III, V

6. Which are acid-base reactions according to the Lowry-Bronsted theory?

$$\text{I.}\quad CH_3\overset{O}{\overset{\|}{C}}O^{\ominus} + CH_3\overset{\oplus}{N}H_3 \rightleftharpoons CH_3\overset{O}{\overset{\|}{C}}OH + CH_3NH_2$$

$$\text{II.}\quad AlCl_3 + Cl^{\ominus} \rightleftharpoons AlCl_4^{\ominus}$$

$$\text{III.}\quad CH_3NH_2 + HCl \rightleftharpoons CH_3\overset{\oplus}{N}H_3\overset{\ominus}{Cl}$$

$$\text{IV.}\quad \overset{\ominus}{C}H_3\overset{\oplus}{Li} + H_2O \rightleftharpoons CH_4 + \overset{\oplus}{Li}\overset{\ominus}{O}H$$

a) I,III b) I, II, III, IV c) I, II, III d) I, III, IV

7. Which of the following are acid-base reactions according to the Bronsted-Lowry theory?

$$\text{I.}\quad (CH_3)_3COH + H_2SO_4 \rightleftharpoons (CH_3)_3C^{\oplus} + H_2O + HSO_4^{\ominus}$$

$$\text{II.}\quad Br_2 + FeBr_3 \rightleftharpoons \overset{\ominus}{Fe}Br_4 + Br^{\oplus}$$

$$\text{III.}\quad CH_3NH_2 + BF_3 \rightleftharpoons CH_3\overset{\oplus}{N}H_2\overset{\ominus}{B}F_3$$

$$\text{IV.}\quad CH_3NH_2 + HCl \rightleftharpoons CH_3\overset{\oplus}{N}H_3\overset{\ominus}{Cl}$$

a) I b) I, III, IV c) II, III d) I, IV

8. Which of the following are acid-base reactions according to the Lewis theory but not according to the Bronsted-Lowry theory?

$$\text{I.}\quad CH_3\overset{O}{\overset{\|}{C}}CH_3 + \overset{\ominus}{C}H_3 \rightleftharpoons CH_3\overset{O^{\ominus}}{\overset{|}{C}}(CH_3)_2$$

$$\text{II.}\quad CH_3CO_2H + H_2O \rightleftharpoons CH_3CO_2^{\ominus} + H_3O^{\oplus}$$

$$\text{III.}\quad NH_3 + BF_3 \rightleftharpoons \overset{\oplus}{N}H_3\overset{\ominus}{B}F_3$$

$$\text{IV.}\quad CH_3OCH_3 + BF_3 \rightleftharpoons (CH_3)_2\overset{\oplus}{O}\overset{\ominus}{B}F_3$$

a) I, II b) III, IV c) I, III, IV d) I, II, III, IV

9. At what percent ionization of acetic acid does the pKa equal the observed pH?

a) 1% b) 10% c) 50% d) 100%

10. What is the percent ionization of acetic acid at pH 2.76 (pKa = 4.76) ?

a) 1% b) 10% c) 50% d) 100%

11. Which of the following acids has the highest pKa?

a) CH_3CO_2H b) H_2O C) NH_4^+ d) HCl

12. What is the pH of a 10% ionized solution of vitamin C (pKa1 = 4.17)?

a) 3.17 b) 5.17 c) 4.17 d) 2.17

13. Arrange the following in the order of increasing basicity (weakest first).

I. OH^- II. Cl^- III. H_2O IV. NH_3

a) II, III, IV, I b) III, I, IV, II c) IV, I, II, III d) III, IV, I, II

14. Arrange the following ions in the order of increasing acidity (weakest first).

I. OH^- II. H_3O^+ III. NH_4^+

a) II, III, I b) I, II, III c) III, II, I d) I, III, II

15. Which of the following ions is the strongest base?

a) $CH_3CH_2O^\ominus$ b) $CH_3CO_2^\ominus$ c) $Cl^\ominus$ d) $CH_3CH_2^\ominus$

16. Arrange the following compounds in the order of increasing acid strength (weakest first).

phenol	pyruvic acid	aspirin	carbonic acid
($K_a = 1.12 \times 10^{-10}$)	($pK_a = 2.49$)	($K_a = 3.3 \times 10^{-4}$)	($pK_a = 6.36$)
I	II	III	IV

a) I, III, IV, II b) I, IV, III, II c) II, IV, III, I d) IV, II, I, II

17. What is the strongest base in the following reaction if the equilibrium constant is much greater than 10?

$$CH{\equiv}CH + \overset{\ominus}{N}H_2 \rightleftharpoons HC{\equiv}\overset{\ominus}{C} + NH_3$$

a) b) c) d)

18. What is the role of diethylether in the following reaction?

$$CH_3CH_2OCH_2CH_3 + BF_3 \rightleftharpoons (CH_3CH_2)_2\overset{\oplus}{O}\,\overset{\ominus}{B}F_3$$

a) Lewis acid b) Lewis base c) Bronsted acid d) Bronsted base

19. What is a reasonable explanation for the following observation?
Acetylene is a stronger acid than ethane.

I. electronegativity II. resonance III. hybridization

a) I, II b) I, III c) II d) I

20. What is a reasonable explanation for the following observation?
H_3O^+ is a stronger acid than NH_4^+.

I. electronegativity II. resonance III. hybridization

a) I, II b) I, III c) II d) I

21. What is a reasonable explanation for the following observation?

Acetic acid is a stronger acid than ethanol.

I. electronegativity II. resonance III. hybridization

a) I, II b) I, III c) II d) I

22. What is the role of water in the following reaction?

$$NH_3 + H_3O^{\oplus} \longrightarrow H_2O + NH_4^{\oplus}$$

a) acid b) base c) conjugate acid d) conjugate base

23. What is the role of methanol in the following reaction?

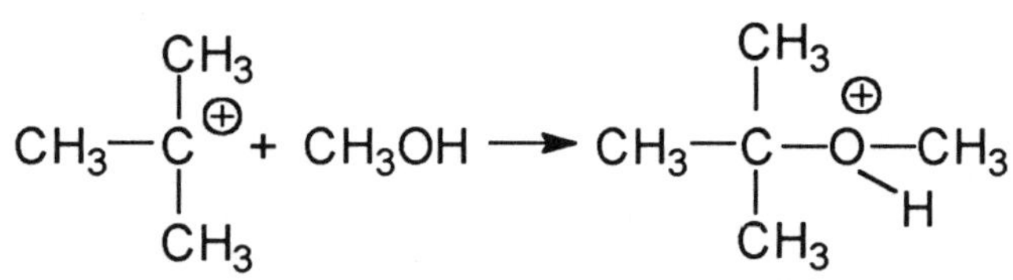

a) Lewis acid b) Lewis base c) Bronsted acid d) Bronsted base

24. Which of the following reactions will proceed extensively to the right?

I. $CH_3CO_2H + NH_3 \rightleftharpoons CH_3CO_2^{\ominus} + NH_4^{\oplus}$

II. $CH_3CO_2^{\ominus} + H_2O \rightleftharpoons CH_3CO_2H + OH^{\ominus}$

III. $NH_4^{\oplus} + OH^{\ominus} \rightleftharpoons NH_3 + H_2O$

IV. $HCN + OH^{\ominus} \rightleftharpoons CN^{\ominus} + H_2O$

a) II, III b) I, IV c) III, IV d) I, III

25. Which of the following equilibria lie considerably toward the left?

I. $CH_3CO_2H + CN^{\ominus} \rightleftharpoons CH_3CO_2^{\ominus} + HCN$

II. $CH_3CO_2^{\ominus} + NH_4^{\oplus} \rightleftharpoons CH_3CO_2H + NH_3$

III. $NH_4^{\oplus} + OH^{\ominus} \rightleftharpoons NH_3 + H_2O$

IV. $CH_3CO_2H + CH_3^{\ominus} \rightleftharpoons CH_3CO_2^{\ominus} + CH_3OH$

a) II, III   b) I, IV   c) III, IV   d) I, III

Answers

1. a)
2. d)
3. d)
4. a)
5. b)
6. d)
7. d)
8. c)
9. c)
10. a)
11. b)
12. a)
13. a)
14. d)
15. d)
16. a)
17. b)
18. b)
19. b)
20. d)
21. a)
22. d)
23. b)
24. b)
25. a)

Chapter 4 Alkenes I

1. What is the correct name for the following structure?

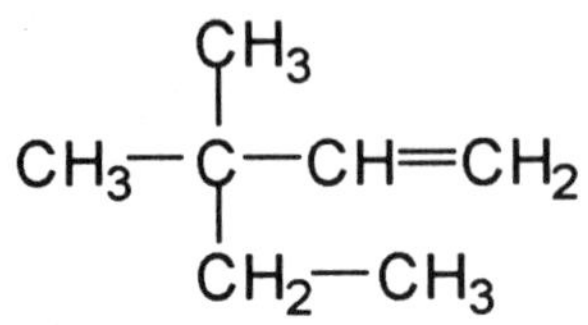

a) 3,3-dimethyl-4-pentene
b) 3-methyl-3-ethyl-1-butene
c) isopropylpentene
d) 3,3-dimethyl-1-pentene

2. What is the correct structure for vinylcyclobutane?

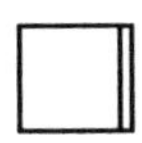

a)

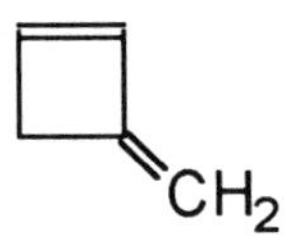

b)

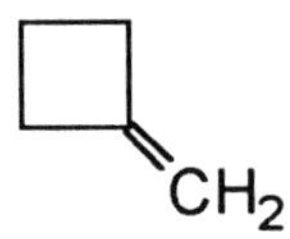

c)

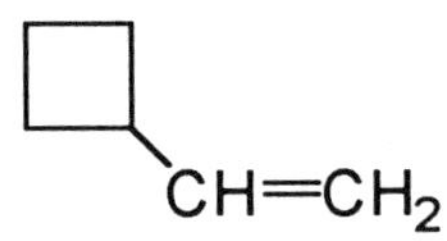

d)

3. What is the correct name for the following structure?

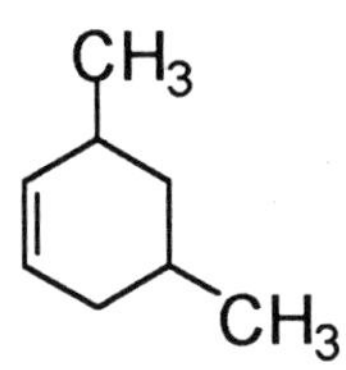

a) 1,3-dimethylcyclohexene
b) 2,4-dimethylcyclohexene
c) 3,5-dimethylcyclohexene
d) 2,4-dimethyl-1-cyclohexene

4. What is the correct name for the following structure?

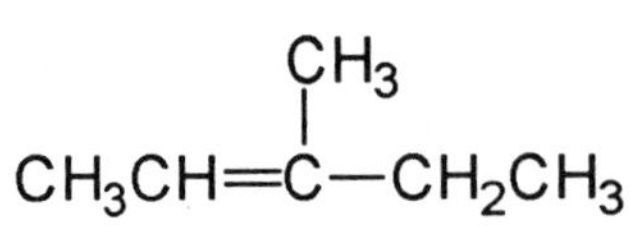

a) 2-ethyl-2-butene
b) 3-methyl-3-pentene
c) 3-methyl-2-pentene
d) 3-ethyl-2-butene

5. What are the approximate carbon-carbon bond angles in allene?

a) 90° b) 109.5° c) 120° d) 180°

6. Arrange the following compounds in the order of increasing bond length for the carbon-carbon single bond (shortest first).

$CH_3{-}CH_3$	$CH{\equiv}C{-}CH_3$	$CH_3{-}CH{=}CH_2$	$CH_2{=}CH{-}CH{=}CH_2$
I	II	III	IV

a I, III, II, IV b) II, IV, III, I c) IV, II, III, I d) I, III, IV, II

7. What is the correct Π* antibonding molecular orbital of ethylene?

a)

b)

c)

d)

8. What is the correct hybridization for the carbon atoms in the following alkene?

$$CH_3{-}C{\equiv}C{-}CH{=}CH_2$$

↑ I (CH₃) ↑ II (C) ↑ III (CH) ↑ IV (CH₂)

a) sp^3, sp, sp^2, sp^2 b) s, p, p, p c) sp^3, sp^2, sp^2, sp^3 d) sp^3, sp, sp, sp

9. Which is the strongest bond in the following molecule?

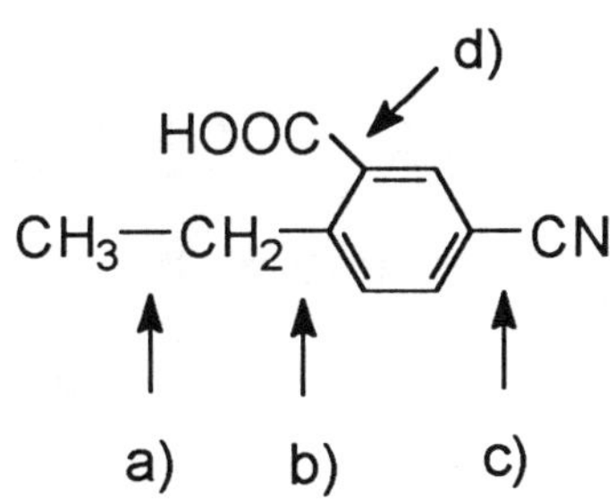

10. Which of the following alkenes do not show cis-trans isomerization?

I. 2-methyl-2-hexene
II. 1-chloro-1-butene
III. 1-methylcyclohexene
IV. 2-methyl-3-hexene

a) I, II b) III, IV c) II, III, IV d) I, III

11. Which of the following alkenes show cis-trans isomerization?

I. 1-chloropropene
II. 3-methylcyclohexene
III. 2,6-dimethyl-2,5-octadiene
IV. 3-ethyl-3-methyl-1-pentene

a) I, II b) II, III c) III, IV d) I, III

12. How many trans isomers are there for an alkene with the formula, C_4H_7Cl ?

a) 2 b) 3 c) 6 d) 8

13. Arrange the following groups in the order of increasing priority (lowest first).

$-CH_2CH_3$	$-CH_2NH_2$	$-CH_2Br$	$-CH_2OH$
I	II	III	IV

a) I, II, IV, III b) III, IV, II, I c) I, IV, II, III d) IV, II, III, I

14. Arrange the following groups in the order of increasing priority (lowest first).

$-CH{=}CH_2$	$-C{\equiv}CH$	$-CN$	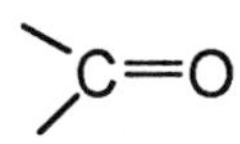
I	II	III	IV

a) IV, III, I, II b) I, II, III, IV c) III, IV, I, II d) III, I, IV, II

15. What is the correct structure for Z-2-bromo-3-methyl-2-pentene?

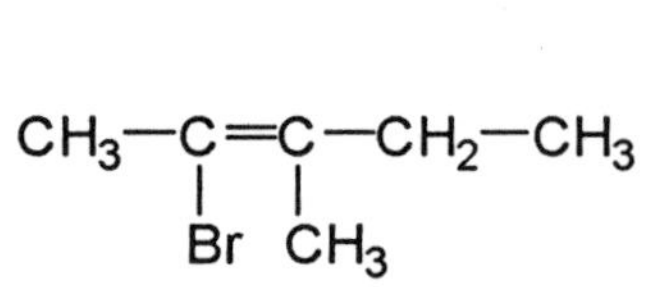
a)

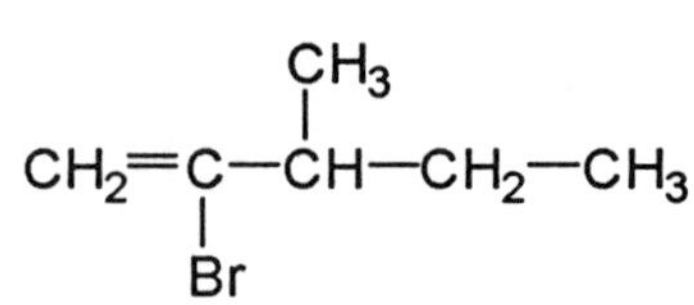
b)

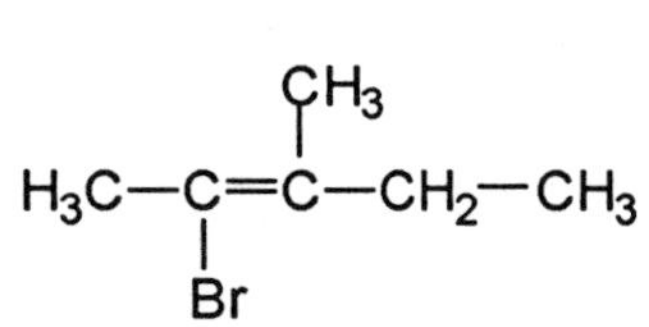
c)

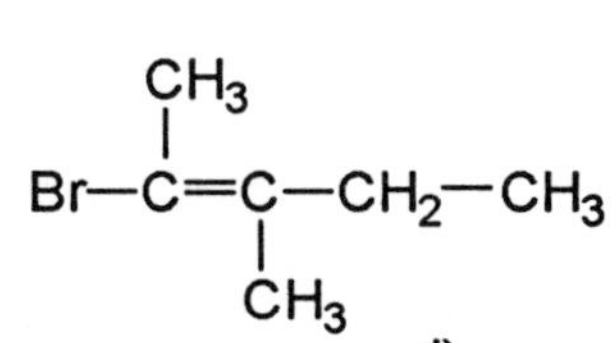
d)

16. Which of the following alkenes are E-forms?

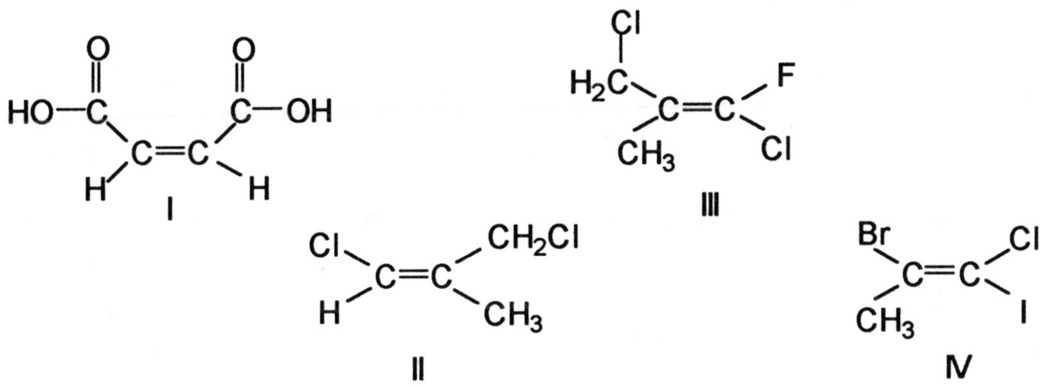

a) I, II b) II, III c) III, IV d) II, IV

17. How many pairs of cis-trans isomers are possible for geraniol?

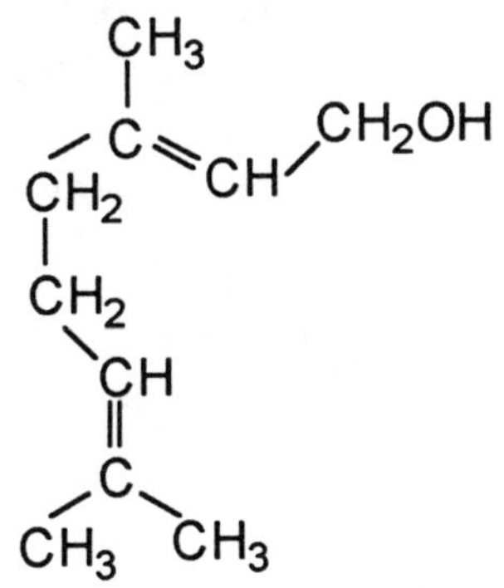

a) 2 b) 3 c) 4 d) 6

18. Which halides give as a major product 2,3-dimethyl-2-pentene on dehydrohalogenation?

I. 1-bromo-2,3-dimethylpentane II. 2-bromo-2,3-dimethylpentane
III. 3-bromo-2,3-dimethylpentane IV. 4-bromo-2,3-dimethylpentane

a) I, II b) II, III c) III, IV d) I, IV

19. Which halides upon dehydrohalogenation give alkenes that do not show cis-trans isomerization?

I. 2-chloropentane II. 2-chlorobutane
III. chlorocyclohexane IV. isobutylchloride

a) I, III b) II, IV c) III, IV d) I, II

20. How many total isomers are possible in the product from dehydrohalogenation of 3-chloro-3-methylhexane?

a) 3　　b) 4　　c) 5　　d) 6

21. What is the product from the dehydrohalogenation of 2-bromonorbornane?

a) norbornane　　b) 1-norbornene　　c) 2-norbornene　　d) 2,5-norbornadiene

22. How many pairs of cis-trans isomers are possible for the product from dehydrohalogenation of 3-bromo-hexane?

a) 1　　b) 2　　c) 3　　d) 4

23. Which halide can give only one product from elimination of HBr?

a) 2-bromo-4,4-dimethylpentane　　b) 2-bromo-3,3-dimethylpentane
c) 3-bromo-2-methylhexane　　d) 3-bromo-2,2-dimethylpentane

24. How many terpene structures can be made from the assembly of 2 isoprene units?

a) 1　　b) 2　　c) 3　　d) 4

25. How many isoprene units are there in vitamin A?

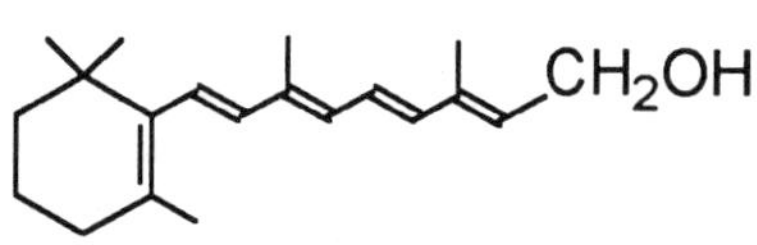

a) 2　　b) 3　　c) 4　　d) 8

Answers

1. d)
2. d)
3. c)
4. c)
5. d)
6. c)
7. b)
8. a)
9. c)
10. d)
11. d)
12. b)
13. a)
14. b)
15. c)
16. c)
17. a)
18. b)
19. c)
20. b)
21. c)
22. b)
23. b)
24. c)
25. c)

Chapter 5 Alkenes II

1. Which alkene has the highest heat of hydrogenation?

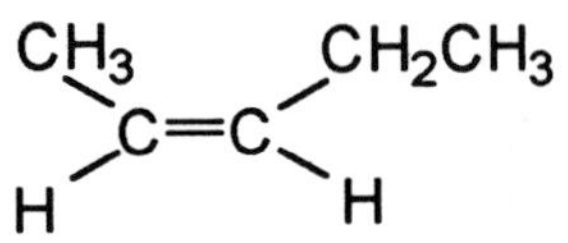

a)

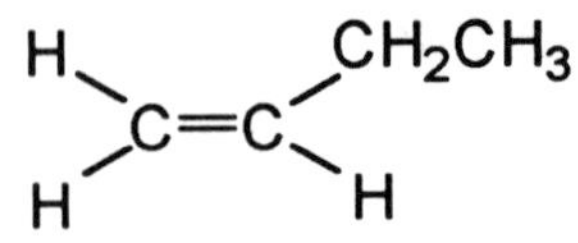

b)

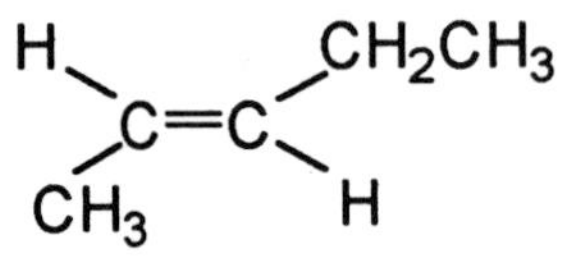

c)

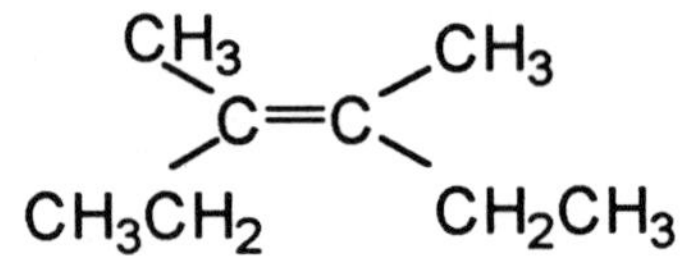

d)

2. Arrange the following isomers of 3-methyloctadiene in the order of increasing stability (lowest first).

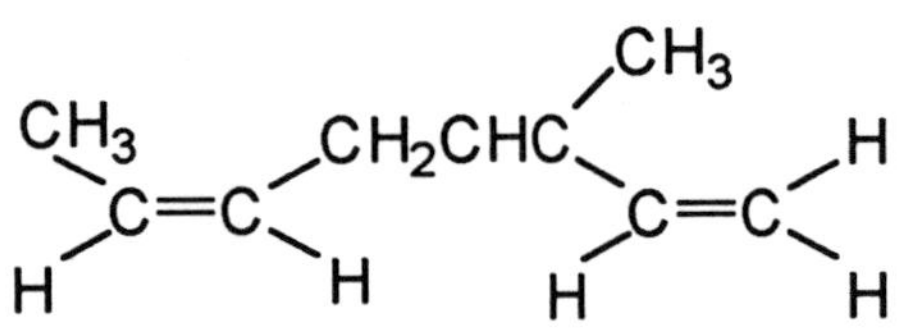

I

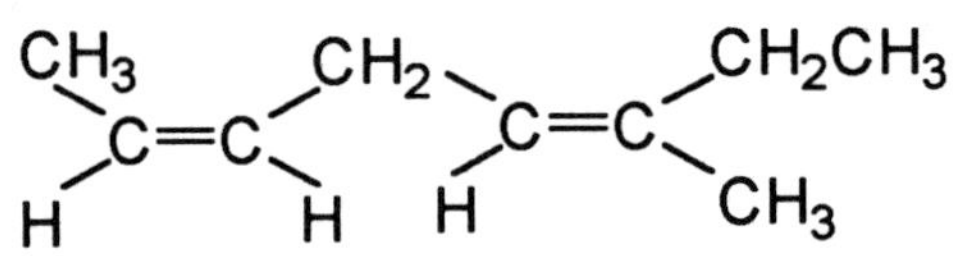

II

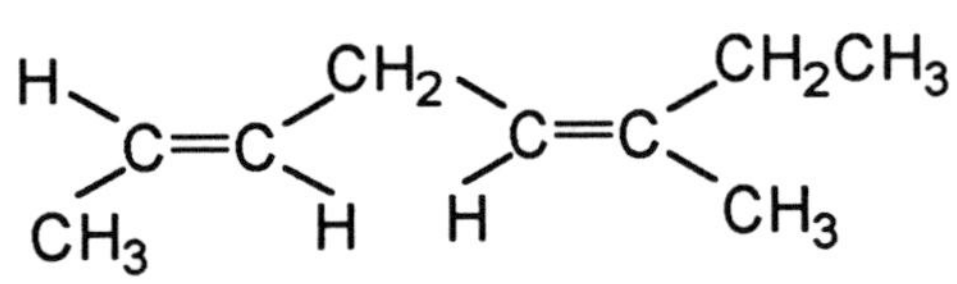

III

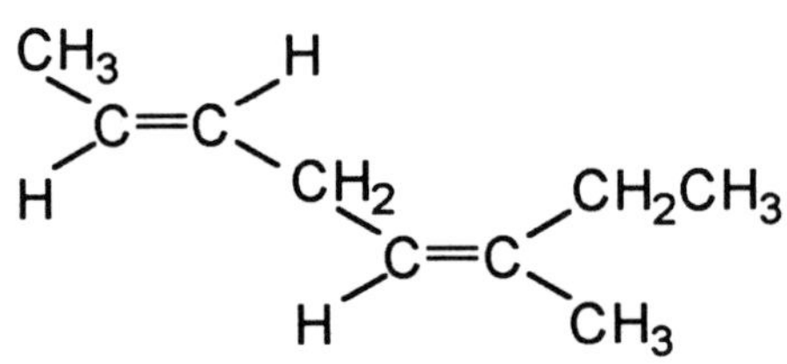

IV

a) I, II, III, IV b) II, III, I, IV c) IV, II, I, III d) II, III, IV, I

3. Indicate which are the allylic hydrogen atom positions in the following molecule.

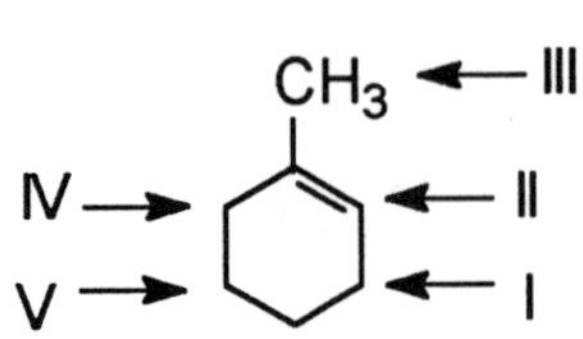

a) II,III b) I,IV c) V d) I, III, IV

4. Arrange the following carbocations in the order of increasing stability (least first).

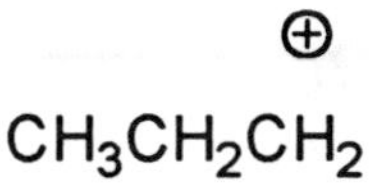 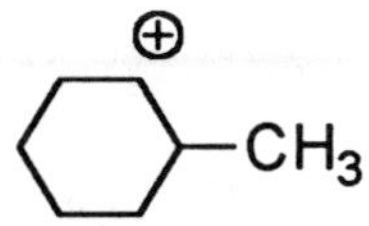 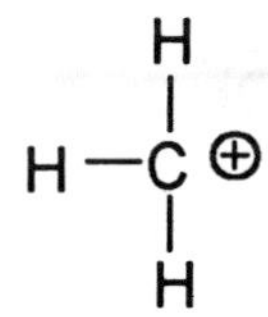 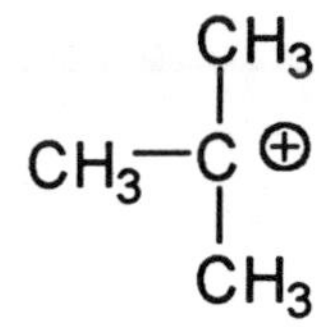

I II III IV

a) I, II, III, IV b) III, I, II, IV c) II, IV, I, III d) I, III, IV, II

5. Which of the following carbocations are likely to rearrange to a more stable intermediate?

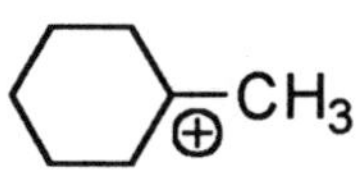 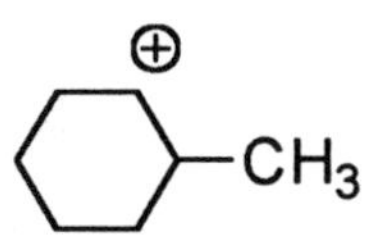 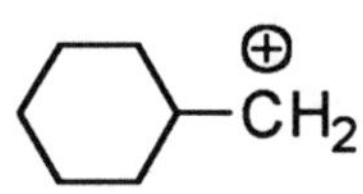 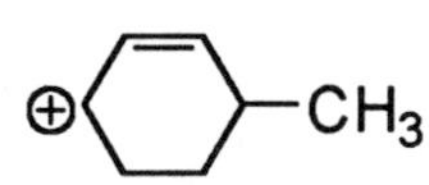

I II III IV

a) II, III b) I, III c) II, IV d) I, IV

6. Which of the following reagents will react with an alkene in an anti addition?

I. Cl_2 II. Br_2 / H_2O III. BH_3 IV. $KMnO_4$

a) I, II b) III, IV c) II, III d) I, IV

7. Which of the following reagents will add to an alkene in a Markovnikov orientation?

I. H_2 II. HBr / peroxides III. HCl IV. $Hg(OAc)_2$ / H_2O

a) I, II, III b) III, IV c) II, IV d) II, III, IV

8. Which of the following react with alkenes by a free radical mechanism?

I. HBr / peroxides II. BH_3 III. $Hg(OAc)_2$ IV. NBS / heat

a) I, II b) III, IV c) I, IV d) II, III

9. According to the mechanism proposed for electrophilic addition of bromine to alkenes, which products do you predict will be formed from the bromination of 1-pentene in water?

I. $CH_3CH_2CH_2\overset{\mathrm{OH}}{\mathrm{C}}HCH_2Br$ II. $CH_3CH_2CH_2\overset{\mathrm{Br}}{\mathrm{C}}HCH_2OH$

III. $CH_3CH_2CH_2\overset{\mathrm{Br}}{\mathrm{C}}HCH_2Br$ IV. $CH_3CH_2CH_2\overset{\mathrm{OH}}{\mathrm{C}}HCH_2OH$

a) I, II b) I, III c) II, IV d) III, IV

10. Which reaction conditions would be used to prepare a vicinal diol from an alkene?

I. H_2O / H_2SO_4 II. cold dilute $KMnO_4$

III. $OsO_4 + H_2O_2$ IV. O_3, $(CH_3)_2S$

a) I, II b) II, III c) III, IV d) II, IV

11. Which of the following reagents add to an alkene by an electrophilic mechanism?

I. Br_2 / H_2O II. Br_2 / CCl_4

III. HBr / peroxides IV. $Hg(OAc)_2$

a) I, II, III b) I, II, IV c) II, IV d) II, III, IV

12. Which of the following reactions involve carbocation intermediates?

a) hydrogenation b) hydroxylation c) hydration d) hydroboration

13. Which of the following are examples of syn addition to alkenes?

I. hydrogenation II. hydration
III. hydrobromination IV. hydroboration

a) I, II b) III, IV c) II, III d) I, IV

14. What is the product of the reaction of 1-methylcyclohexene with BH_3 followed by H_2O_2 / NaOH ?

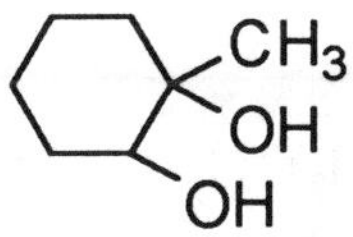

a)

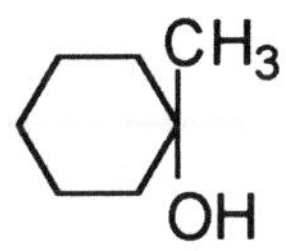

b)

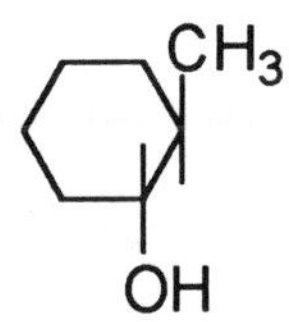

c)

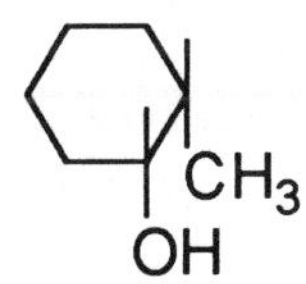

d)

15. What is the major product from acid catalyzed hydration of 2-methyl-2-pentene?

a) 2-methyl-3-pentanol b) 2-methyl-2-pentanol
c) 4-methyl-2-pentanol d) 3-methyl-3-pentanol

16. What is the intermediate for the reaction of propene with hydrogen bromide and hydrogen peroxide?

a) $CH_3-CH(Br)-\dot{C}H_2$ b) $CH_3-\dot{C}H-CH_2Br$ c) $CH_3-CH_2-\dot{C}H_2$ d) $CH_3-\dot{C}H-CH_3$

17. What conditions would you use for the following reaction?

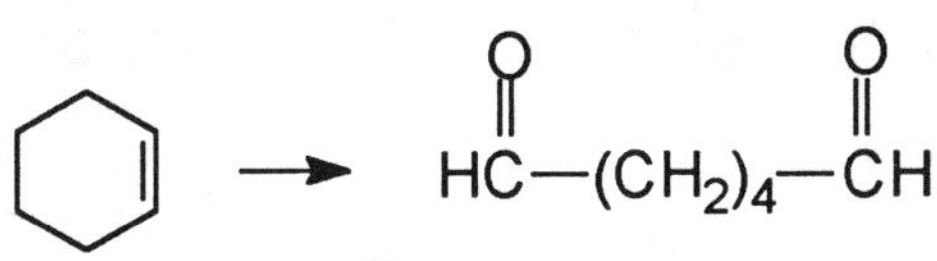

a) $KMnO_4$, H_2SO_4 b) O_3, $(CH_3)_2S$ c) OsO_4, H_2O_2 d) NBS, peroxides

18. What conditions would you use to convert cyclopentene to 3-bromocyclopentene?

a) NBS, peroxides b) HBr, peroxides c) Br_2, CCl_4 d) HBr, H_2SO_4

19. What is the major product from the reaction of cyclopentene with bromine in carbon tetrachloride?

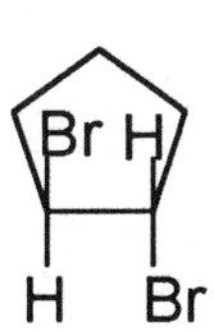

a)

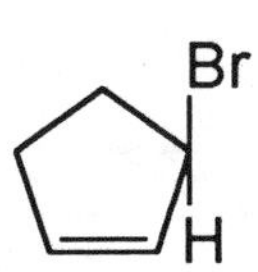

b)

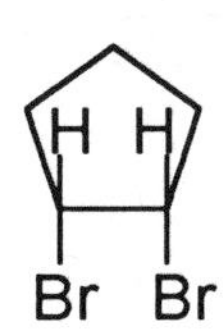

c)

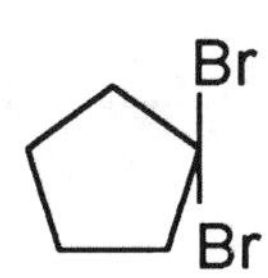

d)

20. What is the major product form the reaction of 1-methylcyclohexene with ozone followed by dimethylsulfide?

a) $CH_3\overset{\overset{O}{\|}}{C}CH_2CH_2CH_2CH_2\overset{\overset{O}{\|}}{C}H$ b) $HO\overset{\overset{O}{\|}}{C}CH_2CH_2CH_2CH_2CH_2\overset{\overset{O}{\|}}{C}OH$

c) $H\overset{\overset{O}{\|}}{C}CH_2CH_2CH_2CH_2CH_2\overset{\overset{O}{\|}}{C}H$ d) $HO\overset{\overset{CH_3}{|}}{C}HCH_2CH_2CH_2CH_2CH_2CH_2OH$

21. Compound A has a molecular formula C_8H_{14} and reacts with hydrogen and platinum to give compound B, C_8H_{16}. What is compound A ?

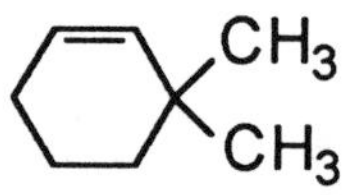

a)

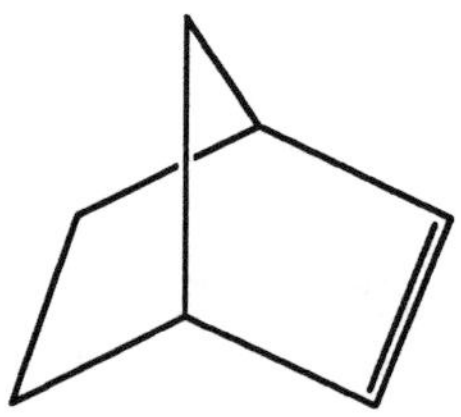
b)

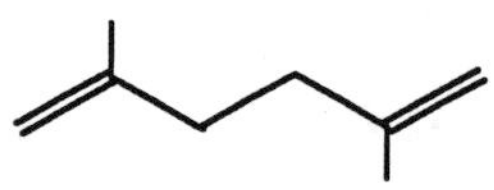
c)

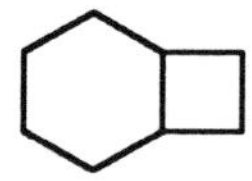
d)

22. What is the major product from the reaction of 1,2-dimethylcyclohexene with deuterium and platinum?

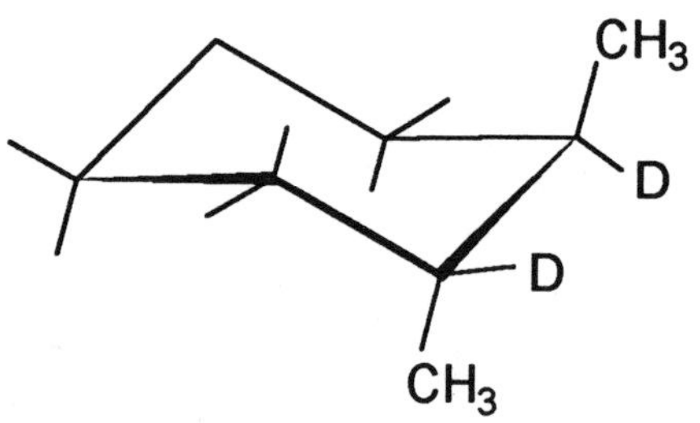

a)

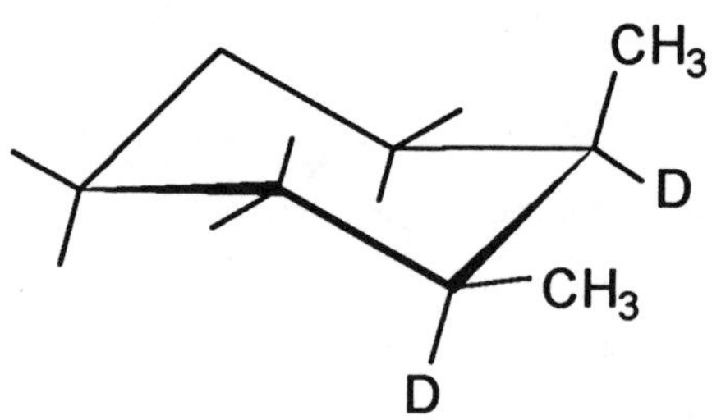

b)

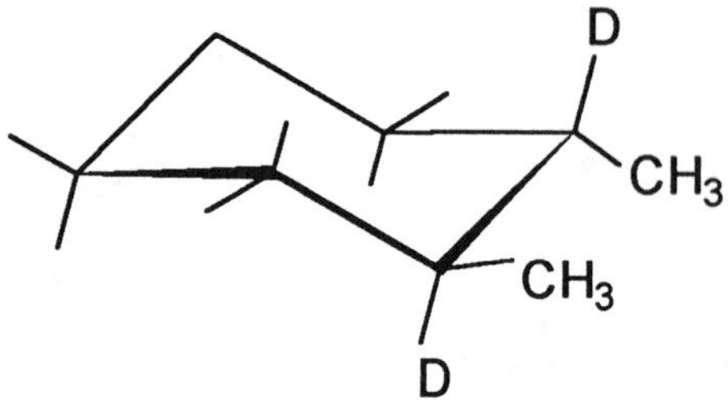

c)

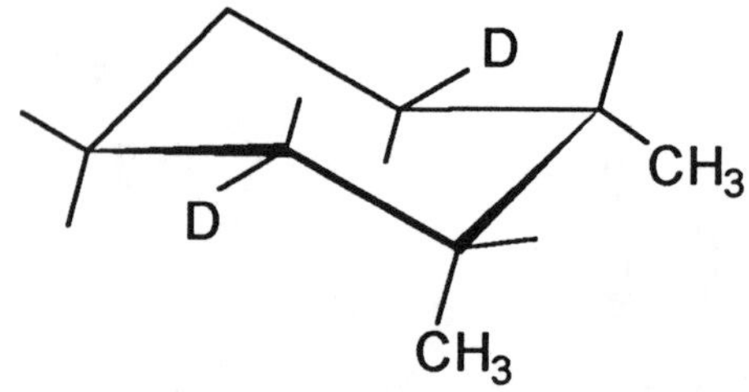

d)

23. Which of the following alkenes will be converted to ketones by reaction with ozone followed by dimethylsulfide?

I. $CH_3CH{=}CH_2$ II. $CH_3CH{=}CHCH_3$

III. $(CH_3)_2C{=}CH_2$ IV. 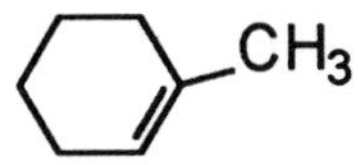

a) I, II b) III, IV c) I, IV d) II, III

24. Bromination of 2-methyl-1-pentene with NBS and peroxides gave a mixture. This mixture reacted with hydrogen with a platinum catalyst to give two isomers. Which of the following are the products?

I. $CH_3CH_2CH_2CH(CH_3)CH_2Br$

II. $(CH_3)_2CHCH_2CH(Br)CH_3$

III. $CH_3C(Br)(CH_3)CH_2CH_2CH_3$

IV. $(CH_3)_2CHCH(Br)CH_2CH_3$

a) I, IV b) I, III c) II, IV d) II, III

25. Which of the following compounds give two isomers when reacted with chlorine in carbon tetrachloride?

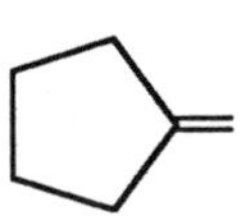

a)

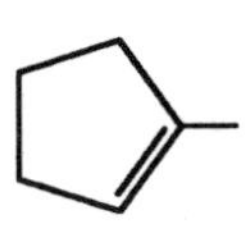

b)

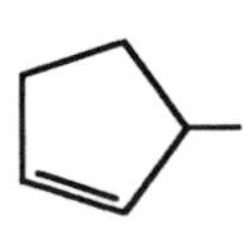

c)

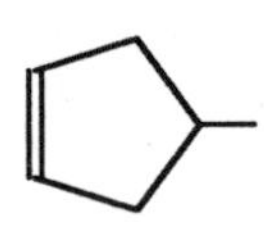

d)

Answers

1. b)
2. a)
3. d)
4. b)
5. a)
6. a)
7. b)
8. c)
9. b)
10. b)
11. b)
12. c)
13. d)
14. c)
15. b)
16. b)
17. b)
18. a)
19. a)
20. a)
21. a)
22. b)
23. b)
24. a)
25. c)

Chapter 6 Alkynes

1. What is the correct IUPAC name for the following structure?

$$CH_3-C\equiv C-CH_2-CH=CH-CH_3$$

a) 5-heptene-2-yne
b) 2-heptene-5-yne
c) 1,5-dimethyl-4-hexen-1-yne
d) 4-hexen-1-yne

2. What is the correct IUPAC name for the following structure?

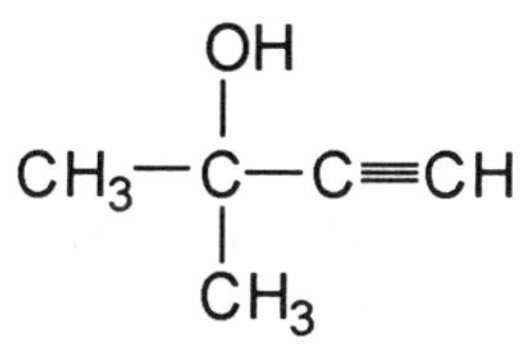

a) 3-butyn-2-methyl-2-ol
b) dimethylhydroxypropyne
c) 2-methyl-2-hydroxy-1-butyne
d) isopropanol acetylene

3. What is the correct IUPAC name for the following structure?

$$CH_3-C\equiv C-CH_3$$

a) dimethyl acetylene
b) 3-butyne
c) 3-butyne
d) 2-butyne

4. What is the correct common name for the following structure?

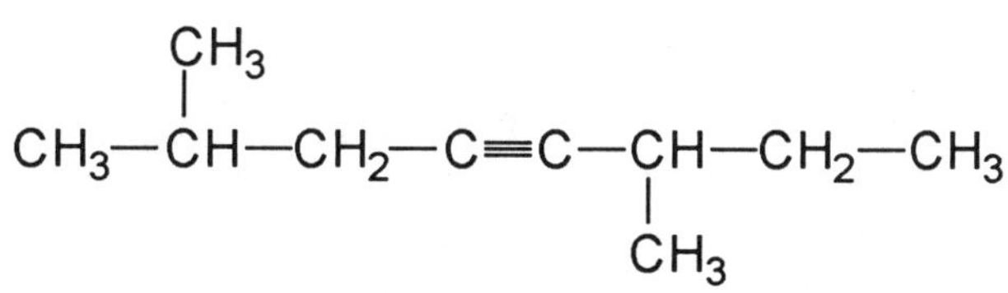

a) isopropyl neopentyl acetylene
b) sec-butyl tert-butyl acetylene
c) isobutyl sec-butyl acetylene
d) diisobutyl acetylene

5. What is the correct common name for the following structure?

$$CH\equiv C-CH=CH_2$$

a) vinyl acetylene
b) ethylene acetylene
c) 1-buten-3-yne
d) 3-buten-1-yne

6. What are the correct orbital hybridizations for the indicated atoms?

$$\underset{\uparrow \atop \text{I}}{CH_3-C}\equiv \underset{\uparrow \atop \text{II}}{C-CH}=\underset{\uparrow \atop \text{III}}{CH-CH_3}$$

a) sp, sp^2, sp^3 b) sp^3, sp, sp^2 c) sp^2, sp, sp^3 d) sp^3, sp^2, sp

7. How many atoms lie in a straight line in 1-pentyne?

a) 2 b) 3 c) 4 d) 5

8. What function does the alkyne suicide substrate play in the inactivation of an enzyme involved in bacterial synthesis of fatty acids?

a) dehydration of an alcohol to form an alkene
b) stereochemical inactivator
c) formation of an allene intermediate that bonds to the enzyme
d) formation of the conjugate base

9. What are the best conditions for preparing 2-hexyne from 2-chlorohexane?

a) $\xrightarrow{NaOCH_3}$ $\xrightarrow{Br_2 / CCl_4}$ $\xrightarrow[NH_3(l)]{NaNH_2}$

b) $\xrightarrow{Cl_2}$ $\xrightarrow[NH_3(l)]{2\ NaNH_2}$

c) $\xrightarrow{NaOH}$ $\xrightarrow{Cl_2}$ $\xrightarrow{2\ Na / NH_3\ (l)}$

d) $\xrightarrow{H_2SO_4}$ $\xrightarrow{HCl}$ $\xrightarrow{NaH}$

10. What are the best conditions for converting ethylene to dideutero acetylene?

a) $\xrightarrow{Cl_2}$ $\xrightarrow[D_2O]{H_2SO_4}$ $\xrightarrow[PdCl_2\ CuCl_2]{CH_3COOD}$

b) $\xrightarrow{Br_2/CCl_4}$ $\xrightarrow[NH_3(l)]{NaNH_2}$ $\xrightarrow[D_2O]{NaH}$

c) $\xrightarrow[Pd\ CuCl_2]{CH_3COOD}$

d) $\xrightarrow{HBr}$ $\xrightarrow{D_2SO_4}$ $\xrightarrow{D_2O}$

11. What is the order of increasing acidity for the following compounds (weakest first)?

I. $CH{\equiv}CH$ II. NH_3 III. CH_3OH IV. H_2O

a) IV, III, II, I b) II, IV, III, I c) I, II, III, IV d) II, I, III, IV

12. Which of the following equilibria are favored to the right?

I. $CH_3CH_2C{\equiv}C^{\ominus} + CH_3CH_3 \rightleftharpoons CH_3CH_2C{\equiv}CH + CH_3\overset{\ominus}{C}H_2$

II. $CH_3C{\equiv}CH + CH_3CH_2CH_2\overset{\ominus}{C}H_2 \rightleftharpoons CH_3C{\equiv}C^{\ominus} + CH_3CH_2CH_2CH_3$

III. $CH_3CH_2C{\equiv}C^{\ominus} + CH_2{=}CH_2 \rightleftharpoons CH_3CH_2C{\equiv}CH + CH_2{=}\overset{\ominus}{C}H$

IV. $CH_3C{\equiv}C^{\ominus} + H_2O \rightleftharpoons CH_3C{\equiv}CH + HO^{\ominus}$

a) I, II b) II, III c) II, IV d) I, III

13. Which of the following is not an equilibrium of tautomers?

a) 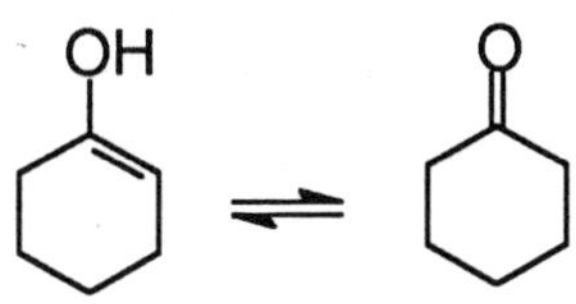

b) $CH_3\overset{O}{\overset{\|}{C}}H \rightleftharpoons CH_2{=}CHOH$

c) $CH_3C{=}CH(OH) \rightleftharpoons CH_3C(OH){=}CH_2$

d) 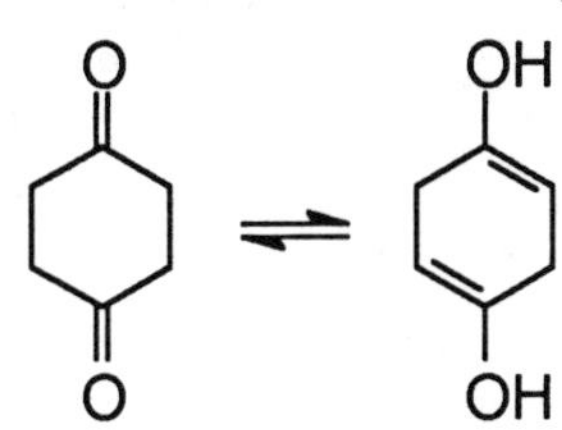

14. Which of the following is a keto-enol equilibria that is favored to the right?

a) 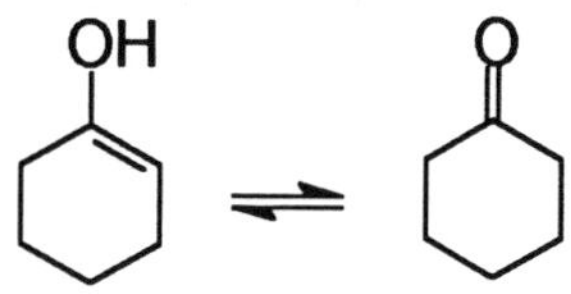 b) 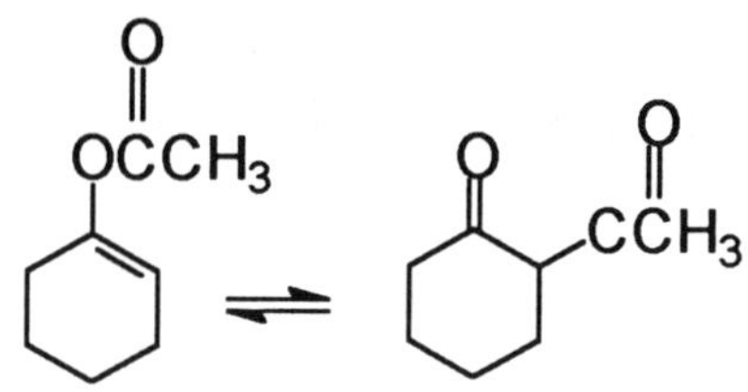

c) $CH_3-\overset{O}{\overset{\|}{C}}-H \rightleftharpoons CH_2{=}CHOH$ d) $CH_3CH{=}\overset{OH}{\overset{|}{C}H} \rightleftharpoons CH_3\overset{OH}{\overset{|}{C}}{=}CH_2$

15. What is the product from the following reactions?

$$CH_3CH{=}CH_2 \xrightarrow[CCl_4]{Br_2} \xrightarrow[NH_3(l)]{NaNH_2} \xrightarrow[2)\ CH_3COOD]{1)\ BH_3}$$

a) 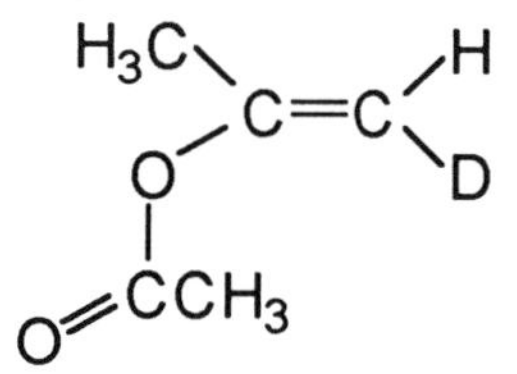 b) 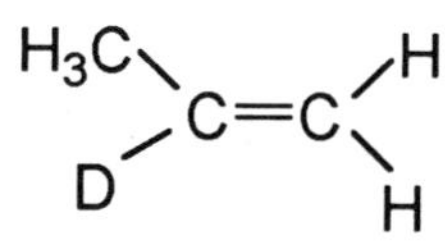

c) 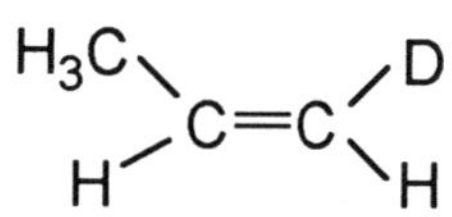 d) 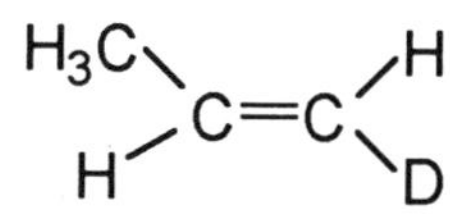

16. What is the product from the following reaction?

$$CH_3C{\equiv}CH \xrightarrow[D_2O]{NaH} \xrightarrow[2)\ CH_3COOH]{1)\ BH_3}$$

a) 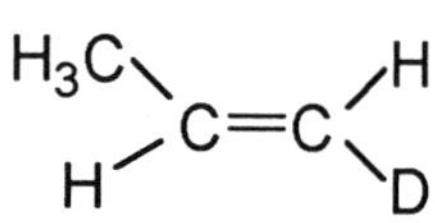 b) 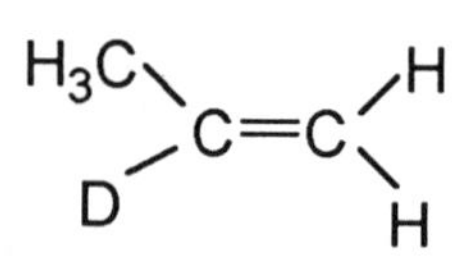

c) 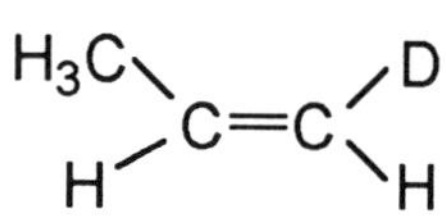 d) 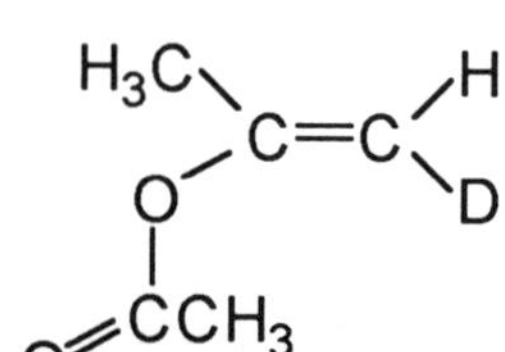

17. What is the product from the following reaction?

$$CH_3C{\equiv}CH + CH_3\overset{\overset{\large O}{\|}}{C}OH \xrightarrow[HgSO_4]{H_2SO_4}$$

a) 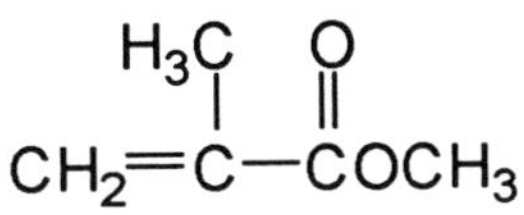

b) 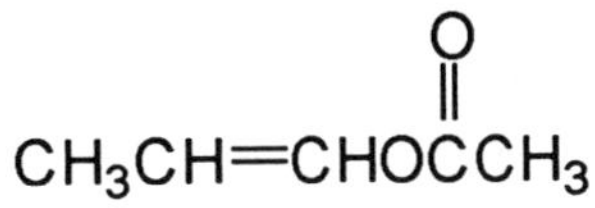

c) 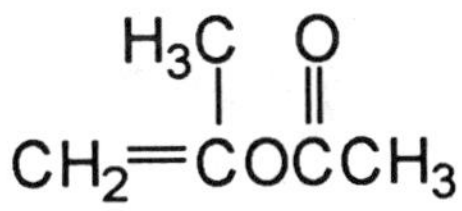

d) 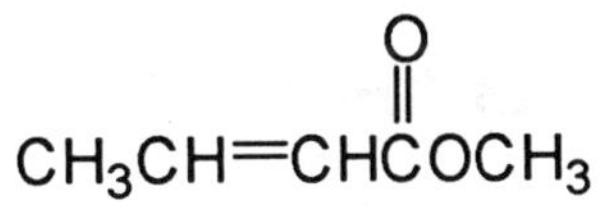

18. What are the best conditions for the following convertion?

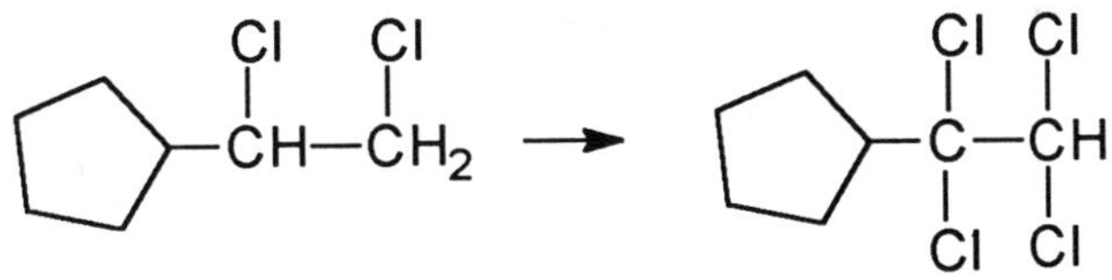

a) $\xrightarrow[NH_3\ (l)]{NaNH_2} \xrightarrow[CCl_4]{Cl_2}$

b) $\xrightarrow[CCl_4]{Cl_2} \xrightarrow[NH_3\ (l)]{NaNH_2}$

c) $\xrightarrow[NH_3\ (l)]{NaNH_2} \xrightarrow{HCl}$

d) $\xrightarrow[CH_3COOH]{Zn} \xrightarrow[CCl_4]{Cl_2}$

19. What are the best conditions for converting 1-pentyne to 1-pentanal?

a) $\xrightarrow{BH_3} \xrightarrow[NaOH]{H_2O_2}$

b) $\xrightarrow{(disia)_2BH} \xrightarrow[NaOH]{H_2O_2}$

c) $\xrightarrow[HgSO_4]{H_2SO_4}$

d) $\xrightarrow[Na]{NH_3\ (l)}$

20. What is the product from the following reaction?

$$CH_3C{\equiv}CH \xrightarrow{HBr} \xrightarrow{HCl}$$

a) 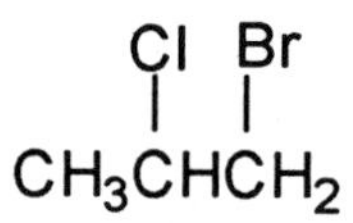 b) 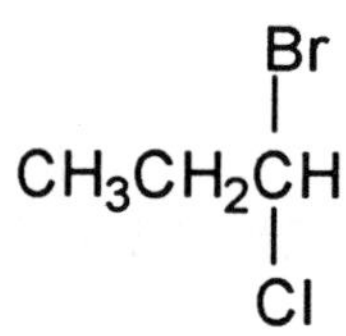 c) 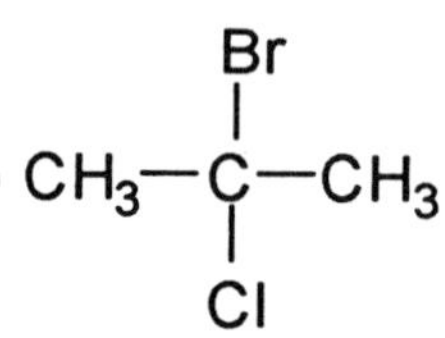 d) 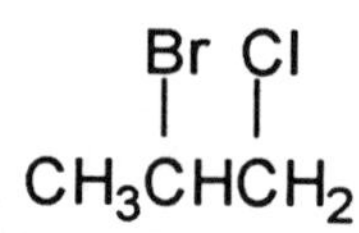

21. What is the product from the reaction of butyne with H_2SO_4 and $HgSO_4$?

a) 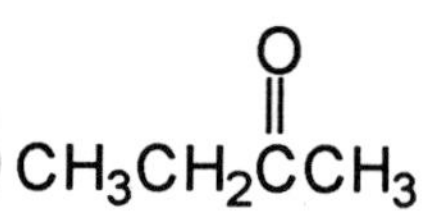 b) CH_3CH_2CHCHO

c) 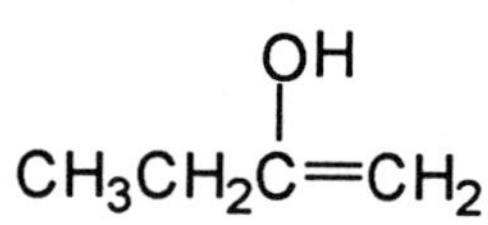 d) $CH_3CH_2CH{=}CH_2$

22. What are the best conditions for converting 3-hexyne to hexane?

a) $Ni + H_2$ b) $Pd / CaCO_3 + H_2$ c) Na / NH_3 d) $NaNH_2 + NH_3(l)$

23. What are the best conditions for converting 3-hexyne to trans-3-hexene?

a) $Ni + H_2$ b) $Pd / CaCO_3 + H_2$ c) Na / NH_3 d) $NaNH_2 + NH_3(l)$

24. What are the best conditions for converting 3-hexyne to cis-3-hexene?

a) $Ni + H_2$ b) $Pd / CaCO_3 + H_2$ c) Na / NH_3 d) $NaNH_2 + NH_3(l)$

25. 3,4-Dibromo-2,5-dimethylhexane was reacted with sodium amide in liquid ammonia to give compound A. Compound A was reacted with potassium permanganate to give compound B. What is a reasonable structure for compound B?

a) $(CH_3)_2\overset{OH}{\overset{|}{C}}H-\overset{OH}{\overset{|}{C}}H(CH_3)_2$ b) $(CH_3)_2CHCHO$

c) $(CH_3)_2\overset{O}{\overset{\|}{C}}-\overset{O}{\overset{\|}{C}}(CH_3)_2$ d) $(CH_3)_2CHCOOH$

Answers

1. b)
2. a)
3. d)
4. c)
5. a)
6. a)
7. c)
8. c)
9. a)
10. b)
11. b)
12. c)
13. c)
14. a)
15. d)
16. c)
17. b)
18. a)
19. b)
20. c)
21. a)
22. a)
23. c)
24. b)
25. b)

Chapter 7 Conjugated Dienes

1. Which of the following are conjugated dienes?

$CH_3-CH=C=CH_2$

I

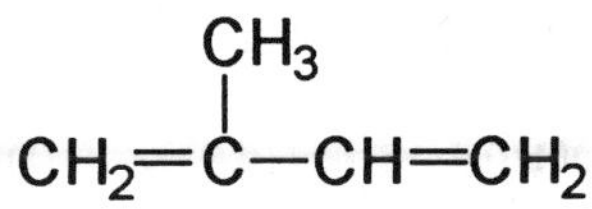

II

III

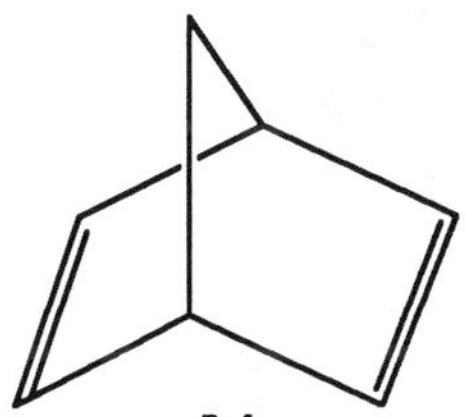

IV

$CH_2=CH-CH_2-CH=CH_2$

V

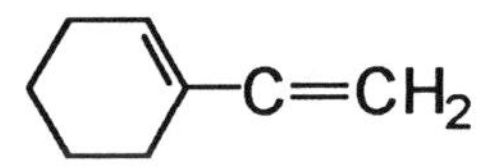

VI

a) I, II, VI b) II, III, VI c) III, IV, VI d) IV, V, VI

2. What is the order of increasing heats of hydrogenation for the following compounds (least first)?

I. 1-pentene II. 1,4-pentadiene
III. cis-2-butene IV. trans-2-butene

a) I, III, IV, II b) III, IV, I, II c) II, IV, III, I d) IV, III, I, II

3. What is the order of decreasing stability for the following compounds (most stable first)?

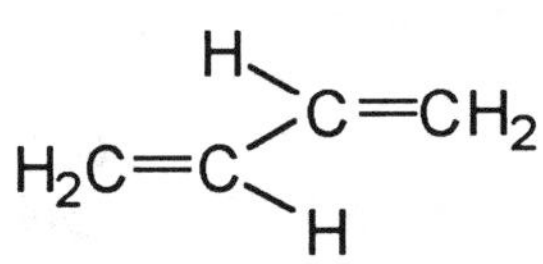

I

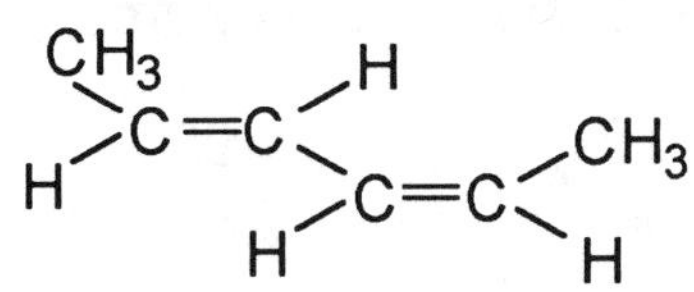

II

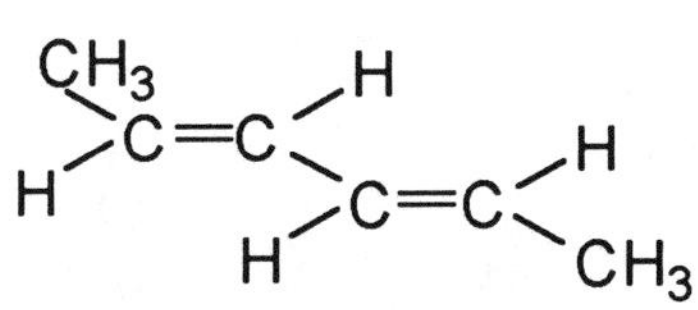

III

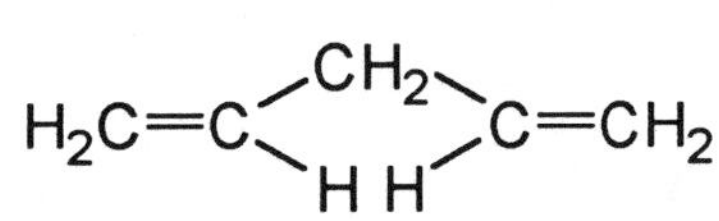

IV

a) I, III, II, IV b) I, II, III, IV c) III, II, I, IV d) III, II, IV, I

4. What is the bonding of the carbon-carbon bonds in 1,3-butadiene?

a) s-s, p b) sp^2-sp^2, Π c) s-p, sp^2-sp^2 d) sp^3-sp^3, p-p

5. Which of these molecular orbital structures represent bonding for 1,3-butadiene?

I

II

III

IV

a) I, II b) II, III c) III, IV d) I, III

6. What is the order of increasing stability for the following cations (least stable first)?

I. $\overset{\oplus}{CH_3}$ II. $CH_2{=}CH{-}\overset{\oplus}{CH_2}$ III. $CH_3CH_2{-}\overset{\oplus}{CH_2}$ IV. $(CH_3)_2\overset{\oplus}{C}{-}CH_3$

a) I, III, IV, II b) IV, III, II, I c) II, I, III, IV d) I, II, III, IV

7. Which of the following are examples of allylic carbocations?

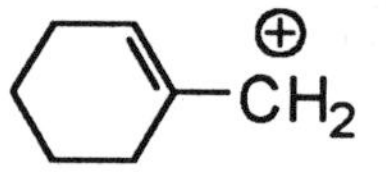

I

$CH_3{-}\overset{\oplus}{CH}{-}CH_2{-}CH{=}CH_2$

II

$CH_2{=}CH{-}\overset{\oplus}{CH}{-}CH_3$

III

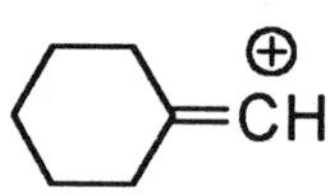

IV

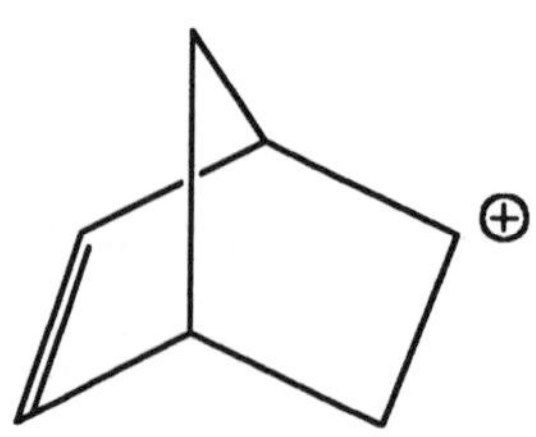

V

$CH_2{=}CH{-}\overset{\oplus}{CH}{-}CH{=}CH_2$

VI

a) II, III, V b) II, III, IV c) I, III, VI d) IV, V, VI

8. What is the major product from 1,2-addition of HBr to 3-methyl-1,3-pentadiene?

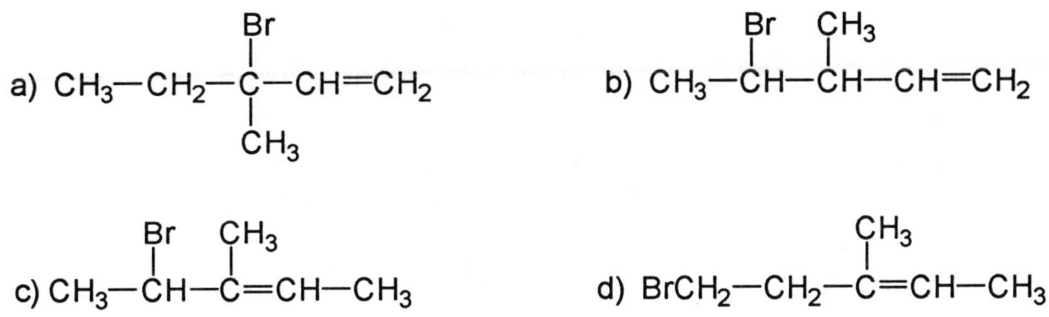

9. What is the major product from the 1,4-addition of HCl to 2-methyl-2,4-hexadiene?

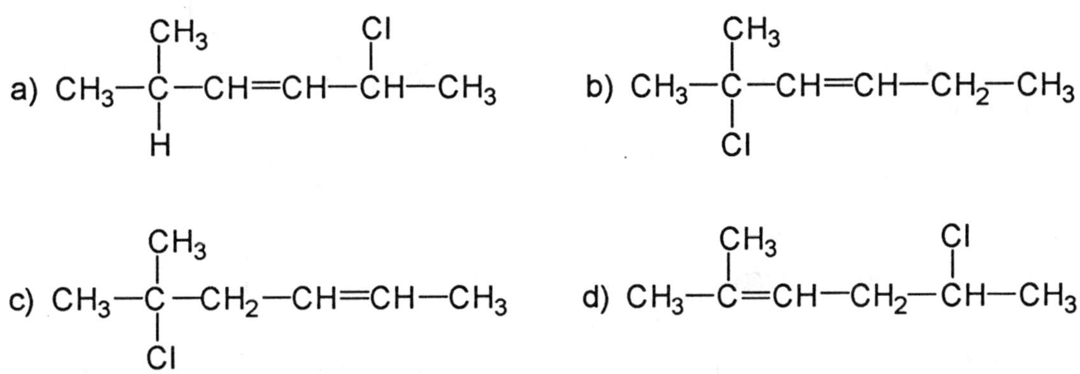

10. What is the principal product from the following reaction?

$$CH_3CH{=}CH{-}CH{=}CHCH_3 + HBr \xrightarrow{-60^{0}\ C}$$

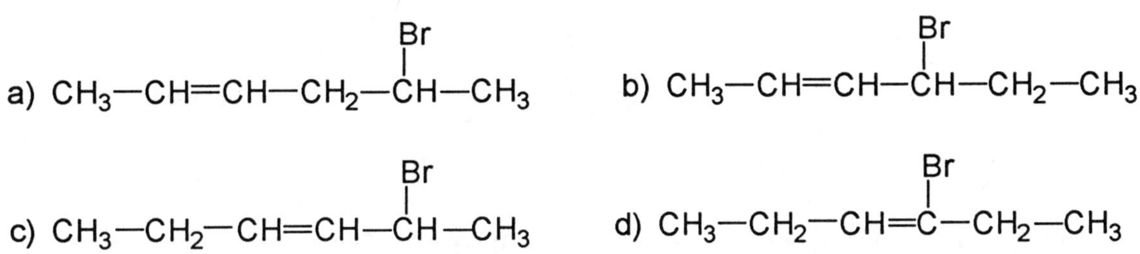

11. What is the principal product from the following reaction?

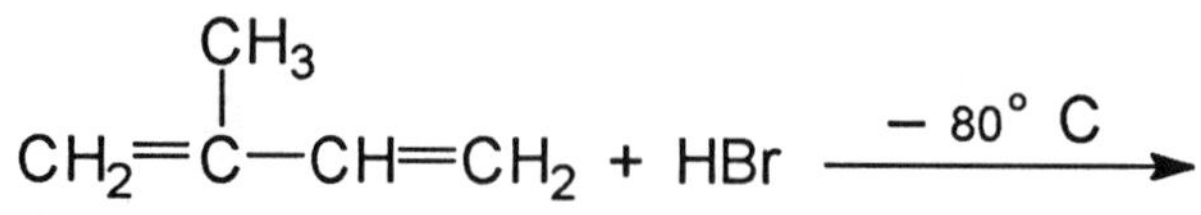

a) 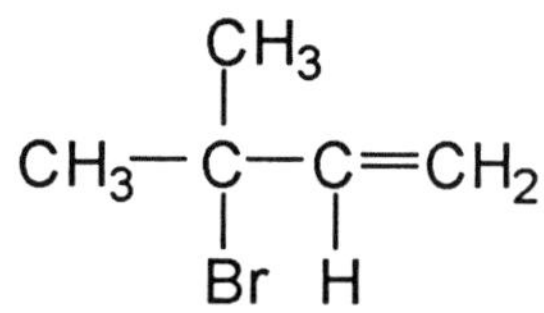

b) 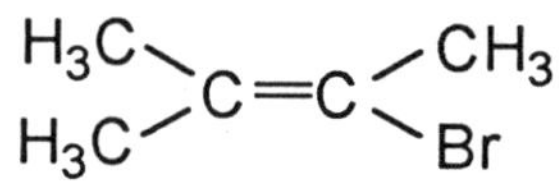

c) 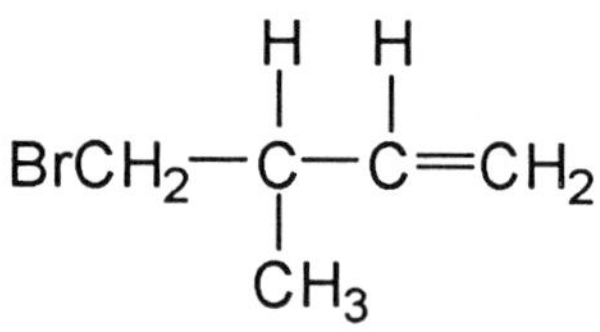

d) 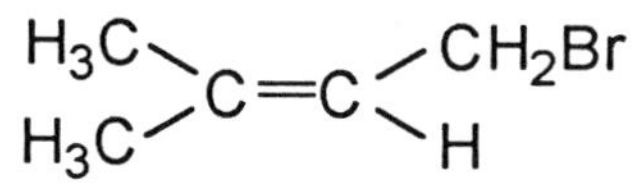

12. Which of the following substituents will activate a diene in a Diels-Alder reaction?

I. $-CH_3$ II. $-CHO$ III. $-C{\equiv}N$

IV. $-C(CH_3)_3$ V. $-OCH_3$ VI. $-NO_2$

a) II, III b) I, IV, V c) II, III, VI d) I, IV

13. Which of the following substituents will activate a dienophile in a Diels-Alder reaction?

I. 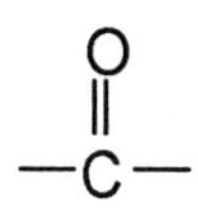 II. 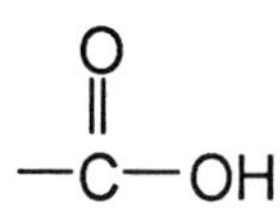 III. 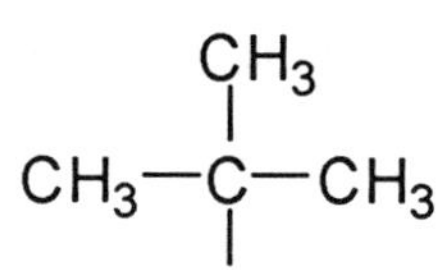

IV. CH_3O- V. 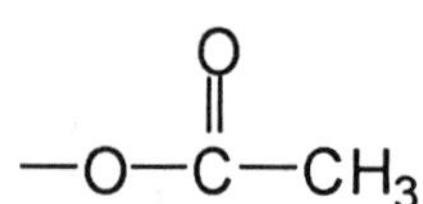 VI. 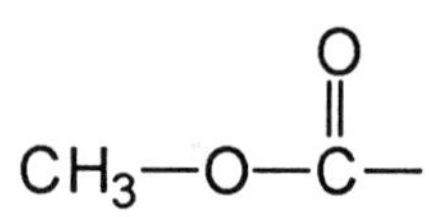

a) I, II, VI b) III, IV, V c) I, II, V, VI d) III, IV

14. Which of the following will react as a diene in a Diels-Alder reaction?

a) 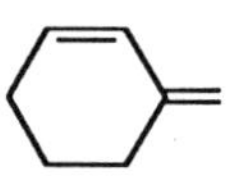 b) 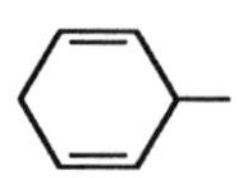 c) 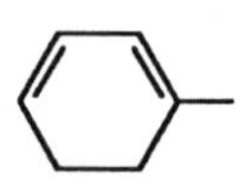 d) 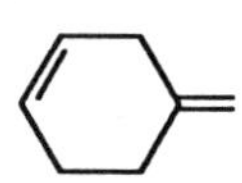

15. What is the product from the Diels-Alder reaction of 1,3-butadiene with 3-buten-2-one?

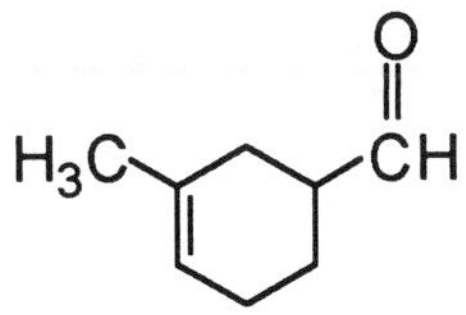
a)

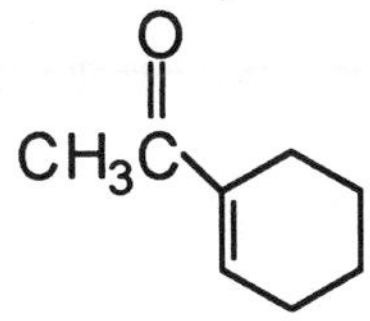
b)

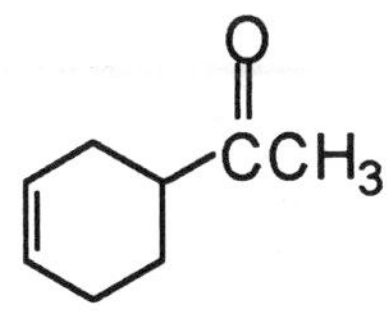
c)

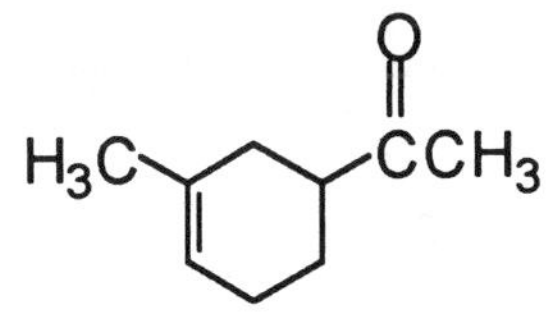
d)

16. What is the product from the following Diels-Alder reaction?

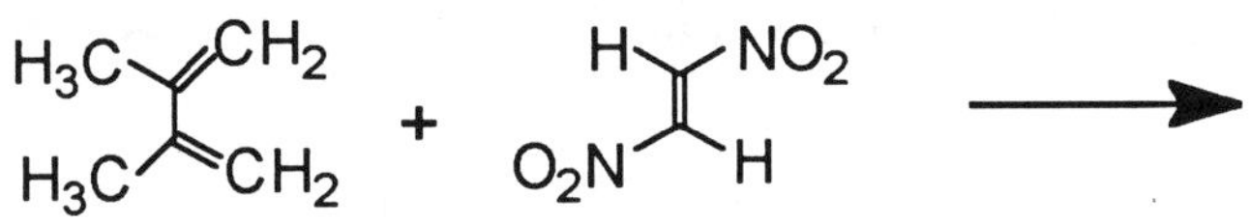

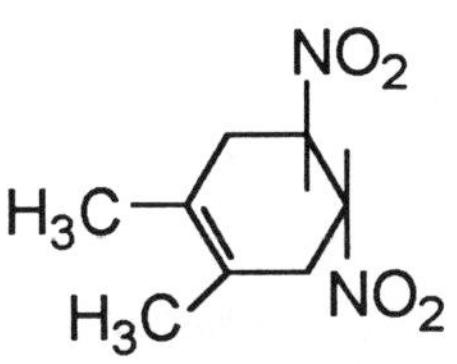
a)

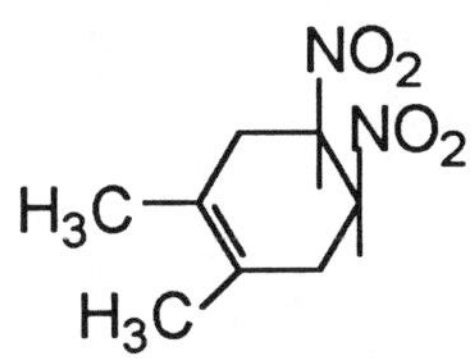
b)

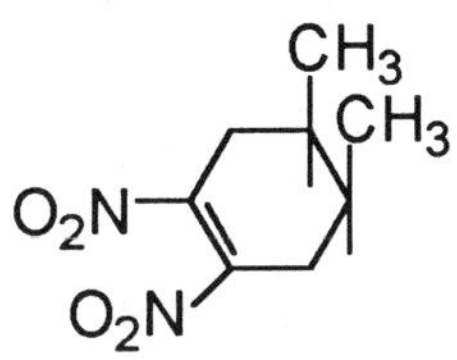
c)

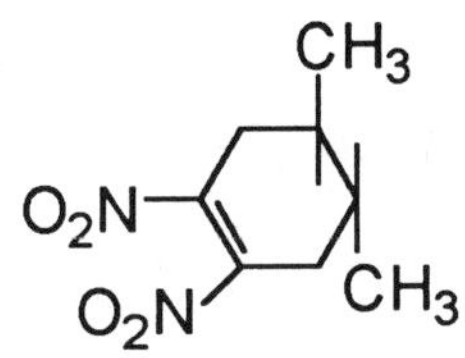
d)

17. What is the product from the following Diels-Alder reaction?

$$CH_3{-}CH{=}CH{-}CH{=}CH_2 \quad + \quad CH_3O_2C{-}C{\equiv}C{-}CO_2CH_3 \longrightarrow$$

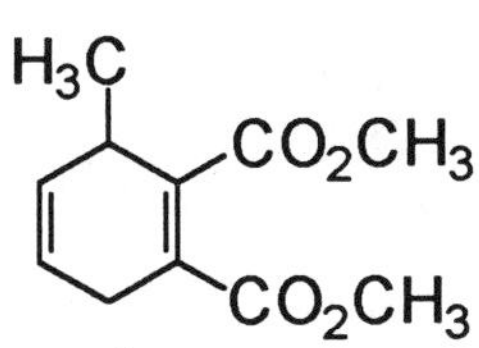
a)

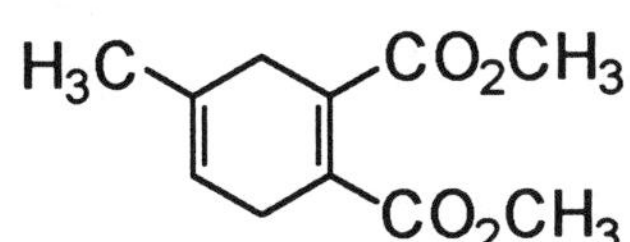
b)

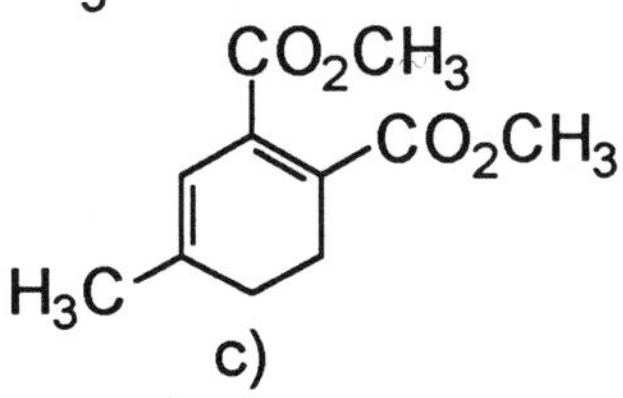
c)

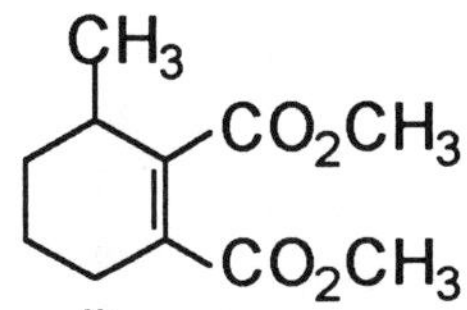
d)

18. What is the product from the following Diels-Alder reaction?

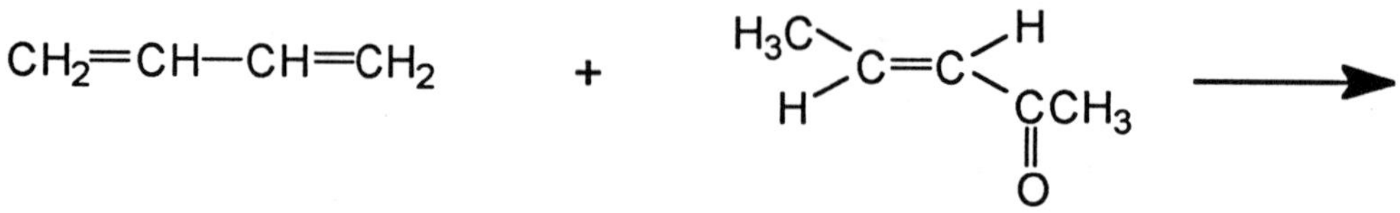

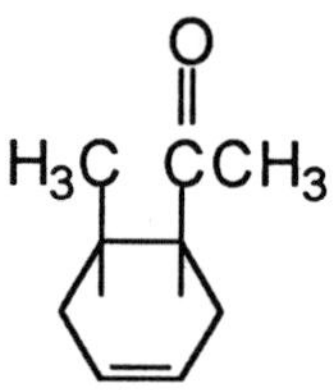

a)

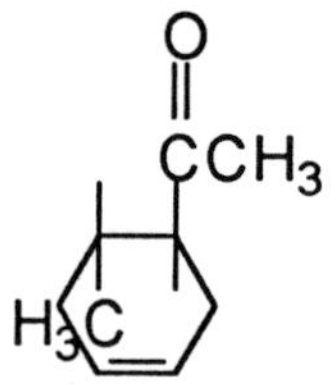

b)

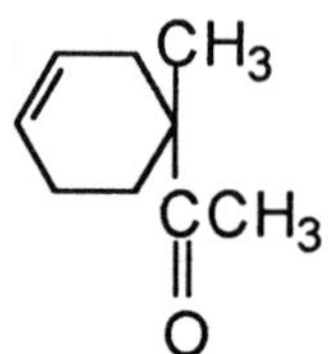

c)

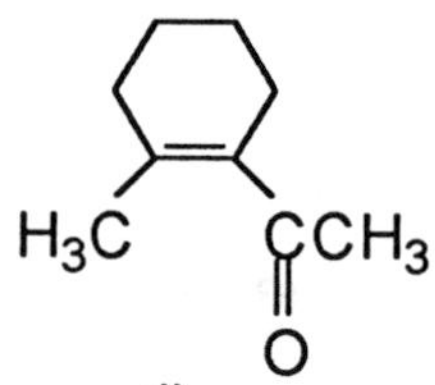

d)

19. What is the product from the following Diels-Alder reaction?

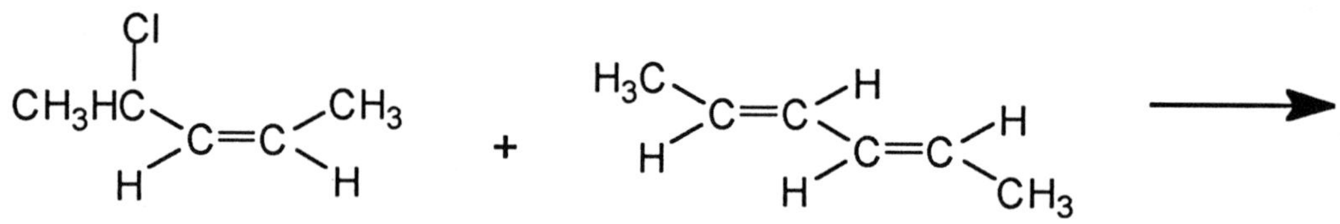

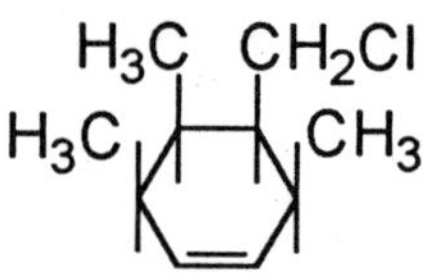

a)

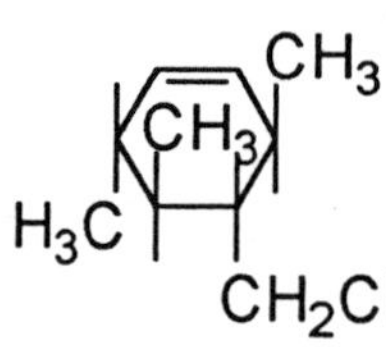

b)

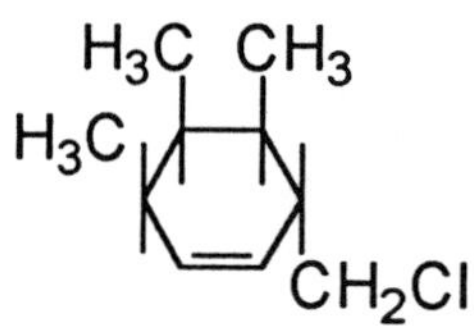

c)

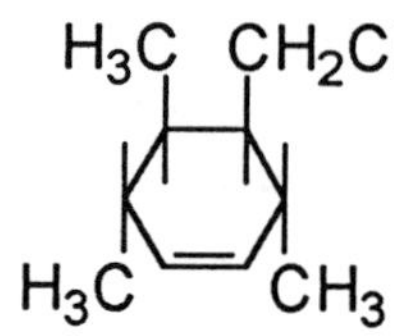

d)

20. What is the Diels-Alder product from the reaction of cyclopentadiene with trans-1,2-dichloroethene?

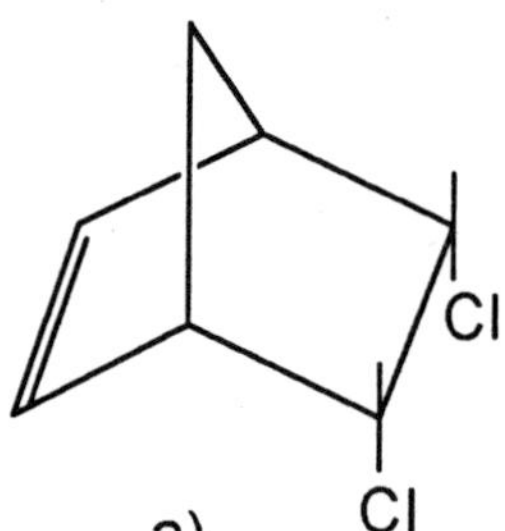

a)

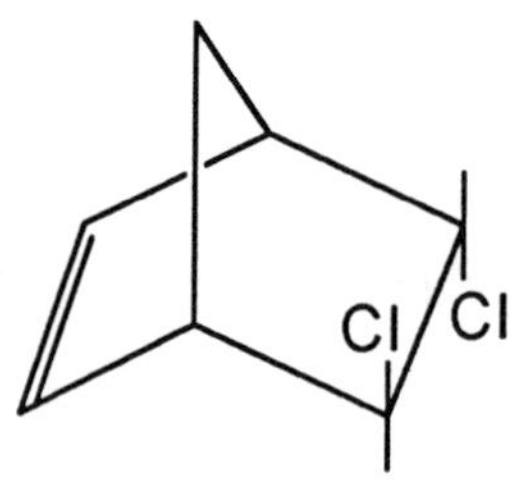

b)

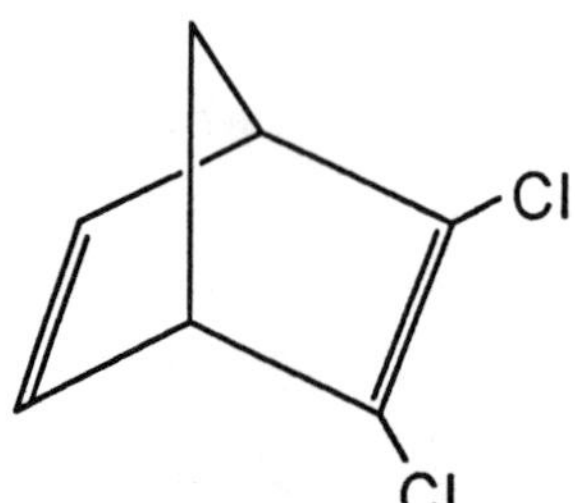

c)

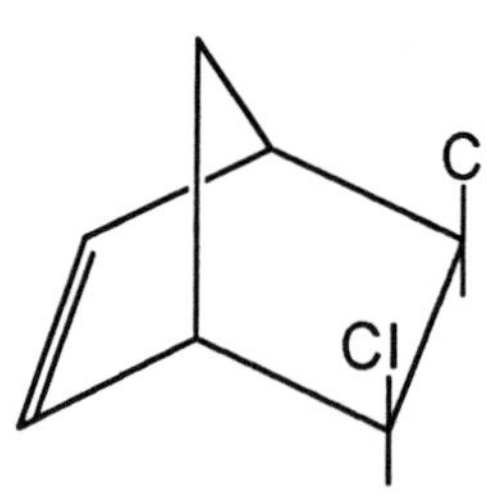

d)

21. What is the Diels-Alder product from two moles of 1,3-cyclohexadiene?

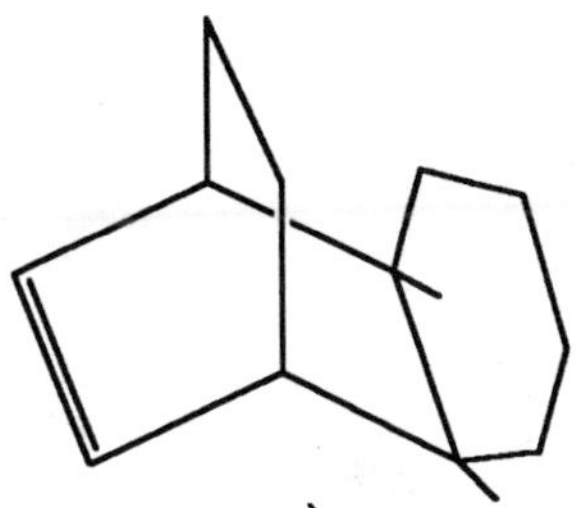

a)

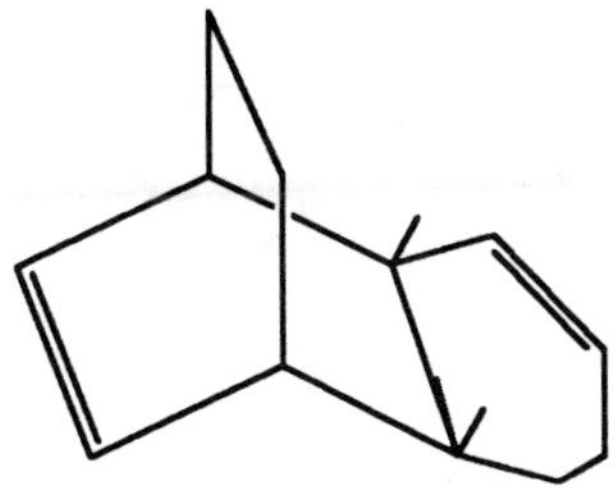

b)

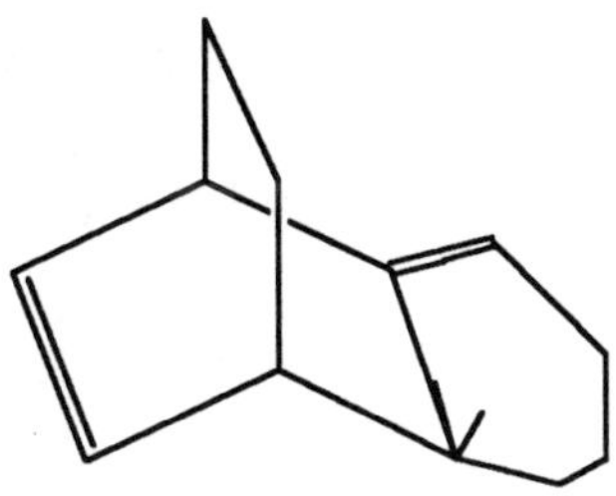

c)

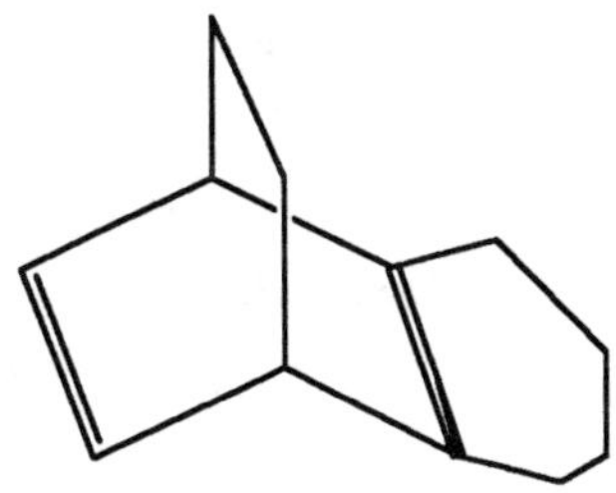

d)

22. Which of the following dienophiles will react with butadiene to give the following product?

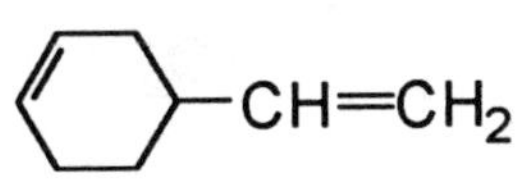

a) $CH_3CH{=}CH{-}CH{=}CH_2$ b) $CH_2{=}CH{-}CH{=}CH_2$

c) $CH_3{-}C{\equiv}CH{-}CH_3$ d) $CH_3\overset{CH_3}{\overset{|}{C}}{=}CH{-}CH_2{-}CH_3$

23. Which of the following dienes will react with 3-penten-2-one to give the following product?

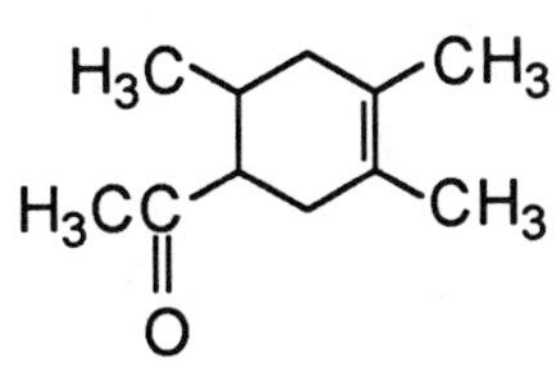

a) $CH_3CH{=}CH{-}CH{=}CH_2$ b) $CH_2{=}CH{-}CH{=}CH_2$

c) $CH_3CH{=}CH{-}CH{=}CHCH_3$ d) $CH_2{=}\underset{CH_3}{\underset{|}{C}}{-}\underset{CH_3}{\underset{|}{C}}{=}CH_2$

24. What combination of diene and dienophile will form the following Diels-Alder product?

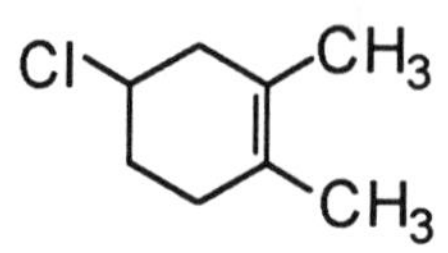

a) 2,3-dimethyl-1,3-butadiene and vinyl chloride
b) 2-methyl-1,3-butadiene and 1-chloropropene
c) 2-chloro-1,3-butadiene and cis -2-butene
d) 1,3-pentadiene and cis-2-chloropropene

25. What combination of diene and dieneophile will react to form the following Diels-Alder product?

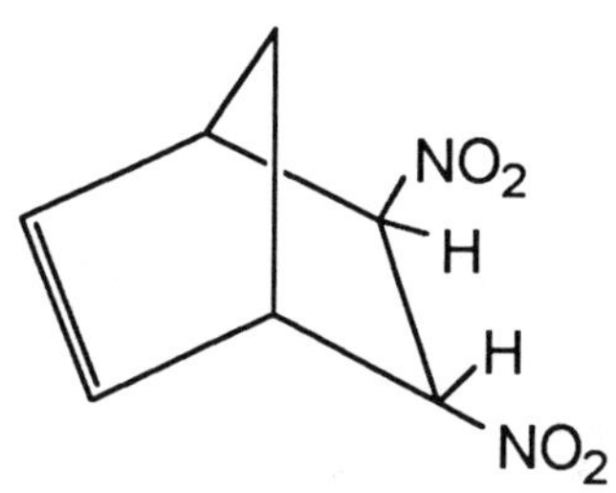

a) 1,3-cyclohexadiene and cis-dinitroethene
b) 1,3-cyclohexadiene and trans-dinitroethene
c) 1,3-cyclopentadiene and trans-dinitoethene
d) 1,3-cyclopentadiene and cis-dinitroethene

Answers

1. b)
2. d)
3. c)
4. b)
5. b)
6. a)
7. c)
8. c)
9. b)
10. b)
11. a)
12. d)
13. a)
14. c)
15. c)
16. a)
17. a)
18. b)
19. a)
20. b)
21. b)
22. b)
23. d)
24. a)
25. c)

Chapter 8 Chirality

1. Which of the following compounds contain stereocenters?

   I. 1-chloropentane II. 2-chloropentane
   III. 3-chloropentane IV. 1,2-dichloropropane

   a) I, II b) III, IV c) I, III d) II, IV

2. Which of the following compounds contain stereocenters?

   I. 2-methylpentane II. chlorocyclohexane
   III. 3-methyl-2-butanol IV. 2-hydroxypropanoic acid

   a) I, II b) III, IV c) I, III d) II, IV

3. Which of the following compounds contain stereocenters?

   I. 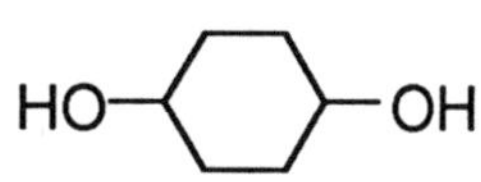
   II. 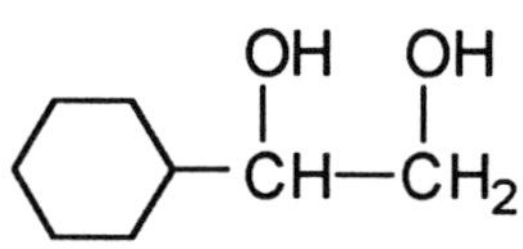
   III. 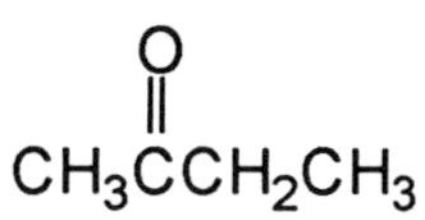
   IV. 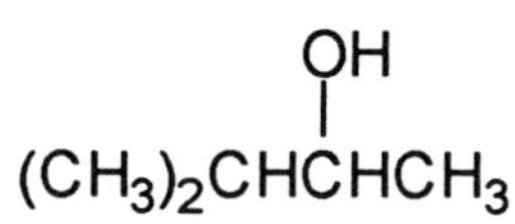

   a) I, II b) III, IV c) I, III d) II, IV

4. Which of the following have multiple stereocenters?

   I. 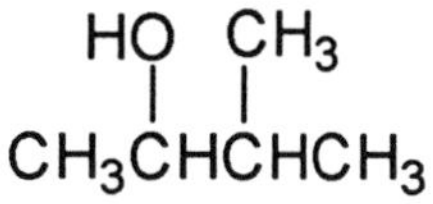
   II. 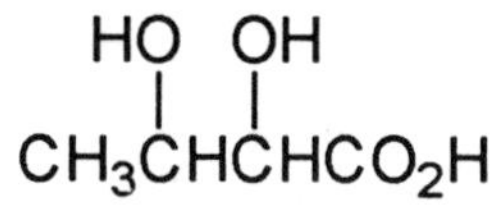
   III. 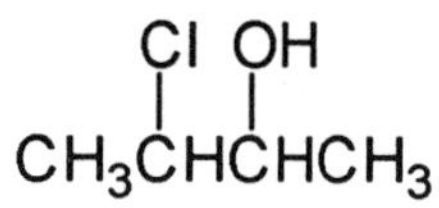
   IV. 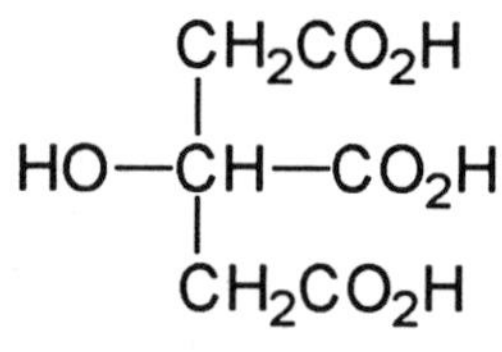

   a) I, II b) III, IV c) I, III d) II, III

5. How many stereoisomers are possible for 1,2-dichlorocyclopentane?

   a) 1 b) 2 c) 3 d) 4

6. How many stereoisomers are possible for 2,3-butanediol?

a) 1 b) 3 c) 2 d) 4

7. How many stereoisomers are possible for the following structure?

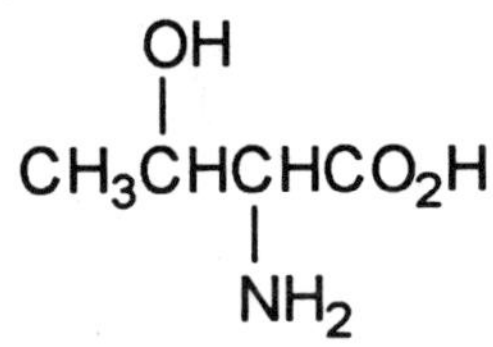

a) 1 b) 2 c) 3 d) 4

8. How many different alkenes with the molecular formula C_5H_9Cl are chiral?

a) 2 b) 3 c) 4 d) 5

9. Which of the following pairs of structures are enantiomers?

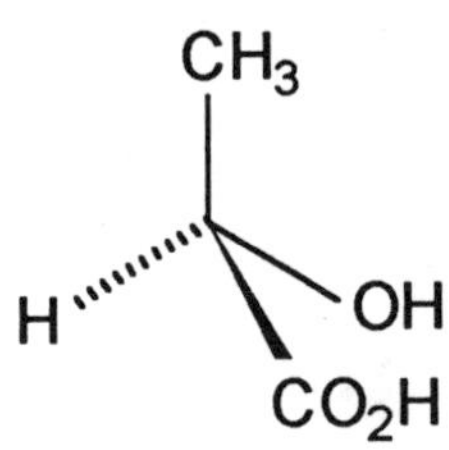
I.
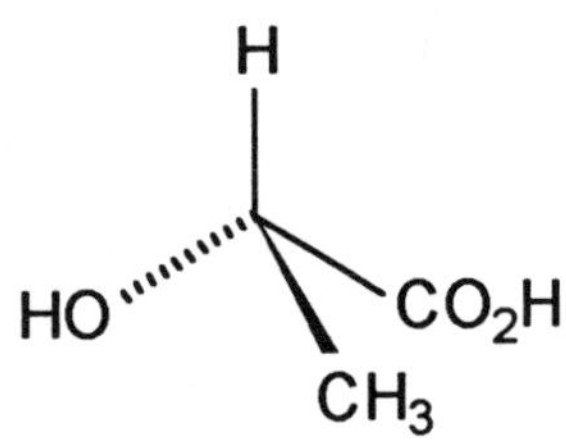

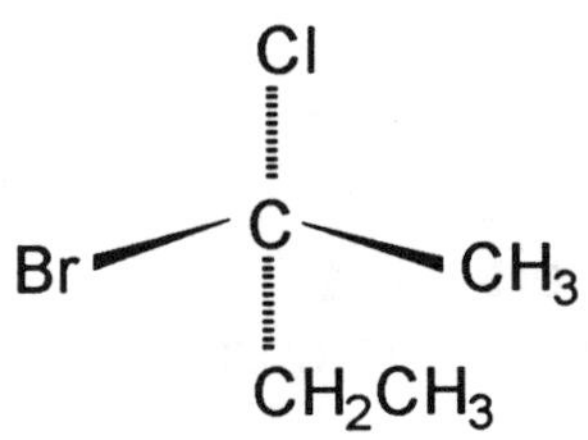
II.
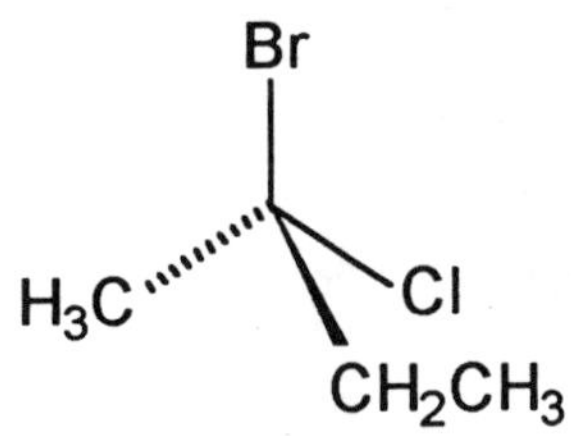

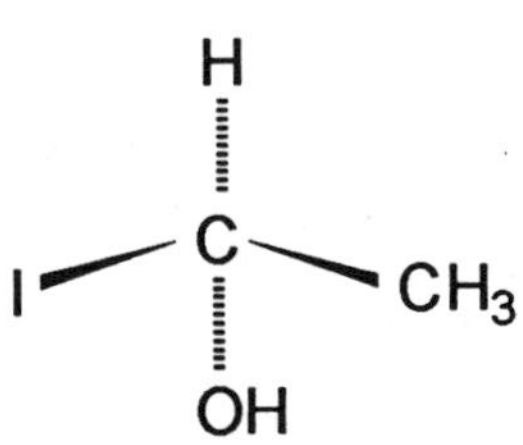
III.
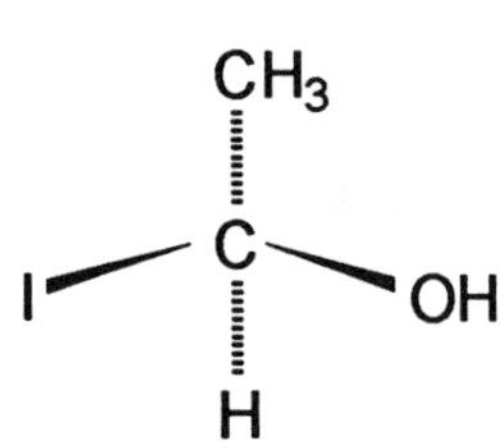

a) I,II b) II, III c) I, III d) I, II, III

10. What is the relationship between these two structures?

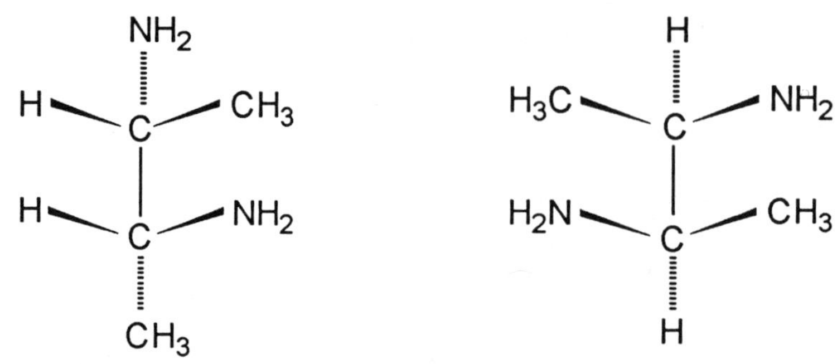

a) identical structures
b) enantiomers
c) diastereomers
d) constitutional isomers

11. What is the relationship between these two structures?

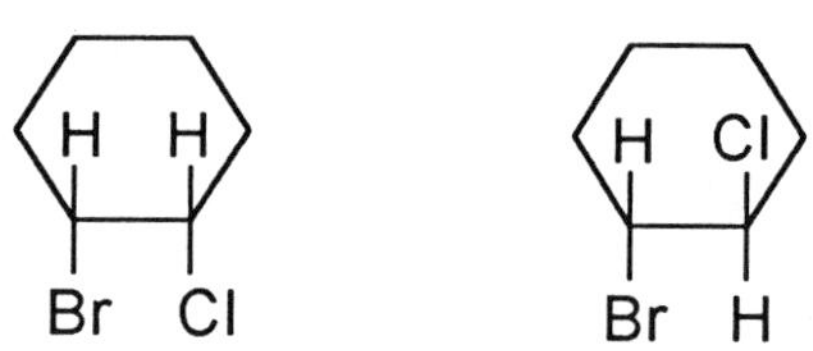

a) identical structures
b) enantiomers
c) diastereomers
d) constitutional isomers

12. Which of the following are meso compounds?

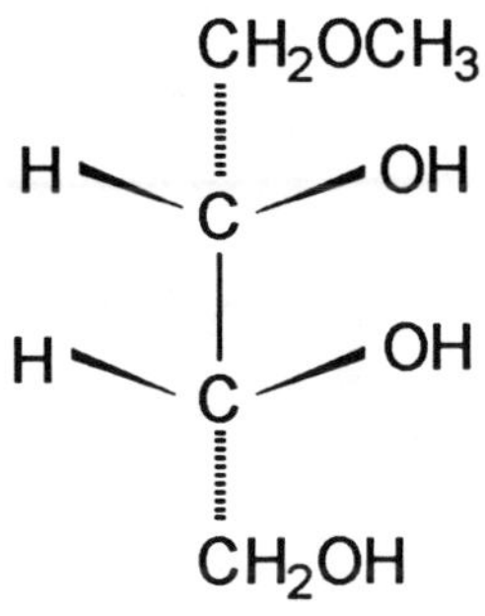

I

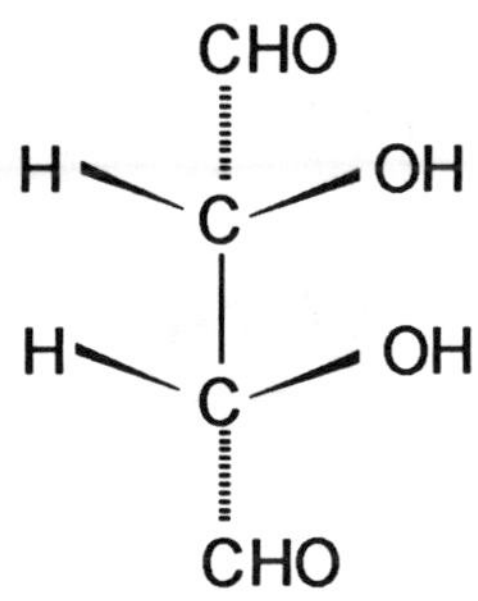

II

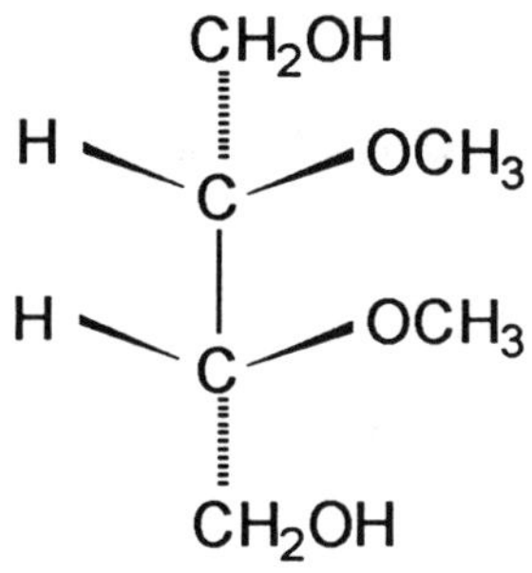

III

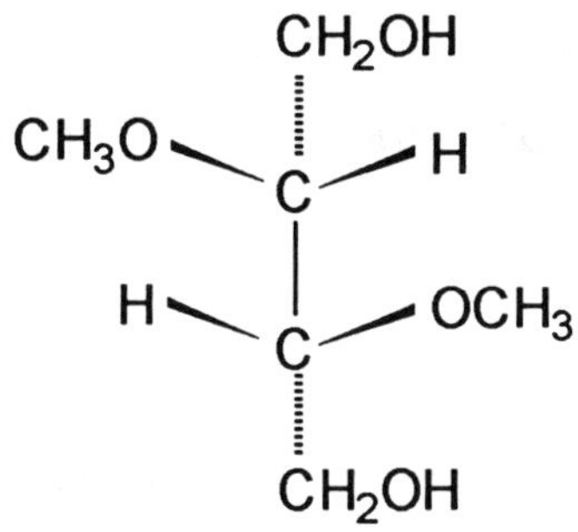

IV

a) I, II   b) I, III   c) II, III   d) III, IV

13. How many stereoisomers are possible for the following structure?

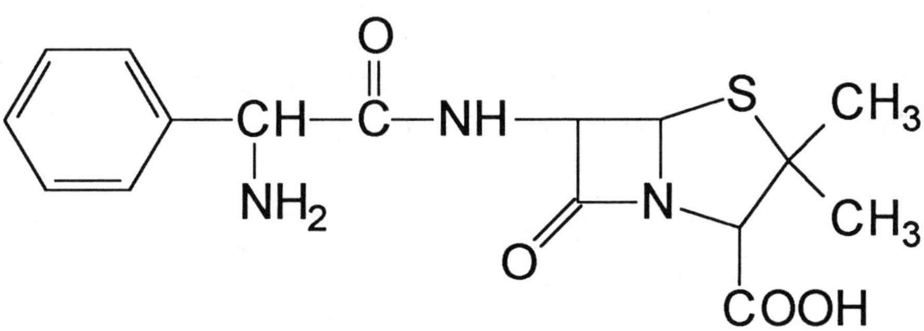

a) 4   b) 8   c) 16   d) 32

14. Which forms of lactic acid are R forms?

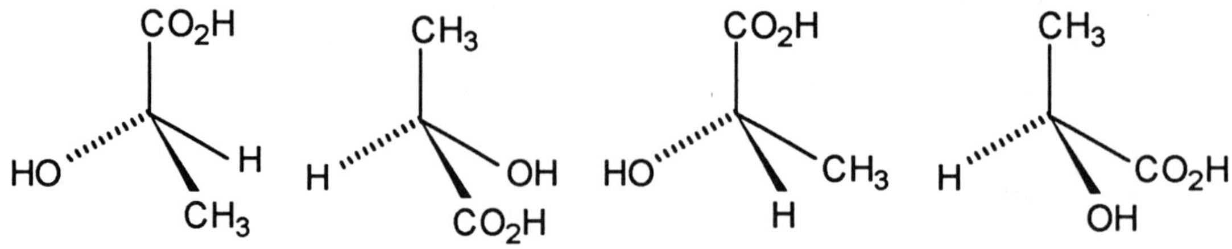

I II III IV

a) I, II b) III, IV c) I, III d) II, IV

15. What is the R,S configuration for the following structure of isocitric acid?

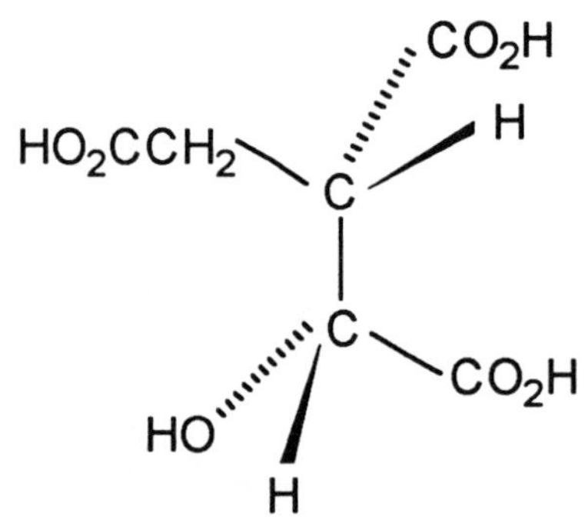

a) 2-R, 3-R b) 2-R, 3-S c) 2-S, 3-R d) 2-S, 3-S

16. For the following reaction, the product, 2,3-dibromohexane, has X possible stereoisomers and only Y are actually formed. What are X and Y?

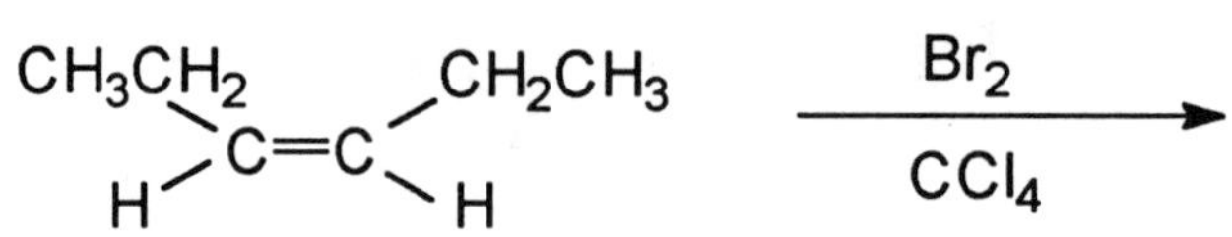

	X	Y
a)	3	1
b)	3	2
c)	4	2
d)	4	3

17. For the following reaction, the product, 2,3-dibromobutane, has X possible stereoisomers and only Y are actually formed. What are X and Y?

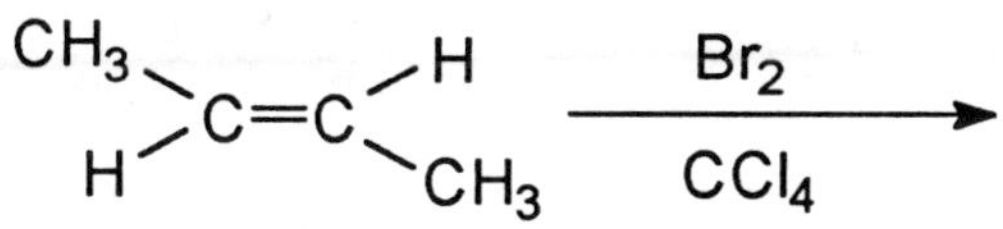

	X	Y
a)	3	1
b)	3	2
c)	4	2
d)	4	3

18. For the following reaction, the product, 1-chloro-1,2-dibromopropane, has X possible stereoisomers and only Y are actually formed. What are X and Y?

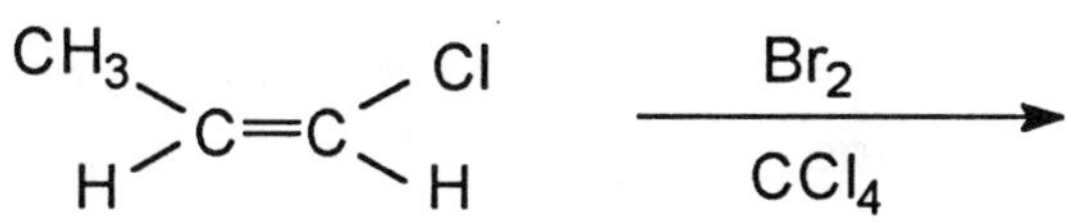

	X	Y
a)	3	1
b)	3	2
c)	4	2
d)	4	3

19. Which of the following reactions leads to a product that is achiral?

a) 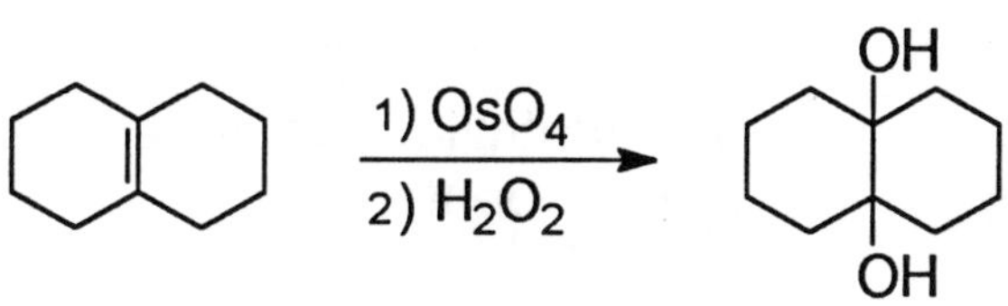

b) 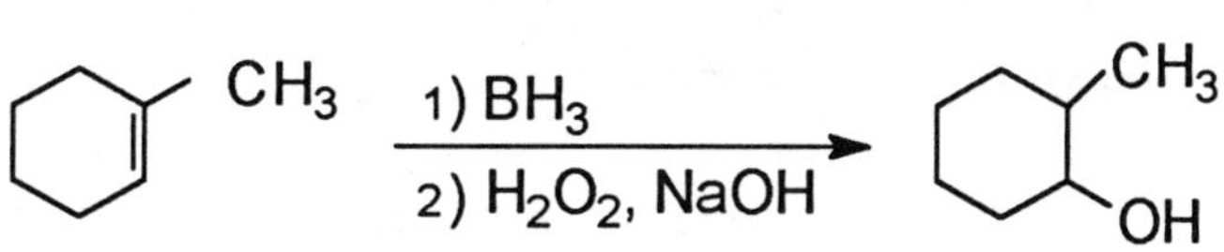

c) 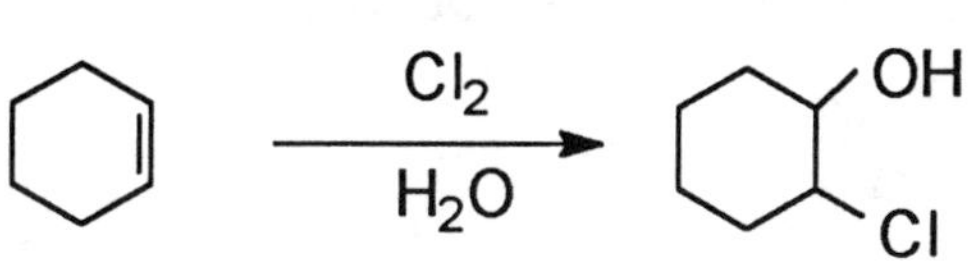

d) 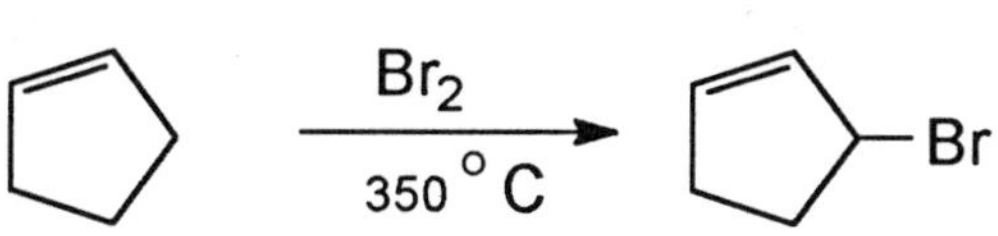

20. Which of the following sets of reaction conditions will lead to a chiral product?

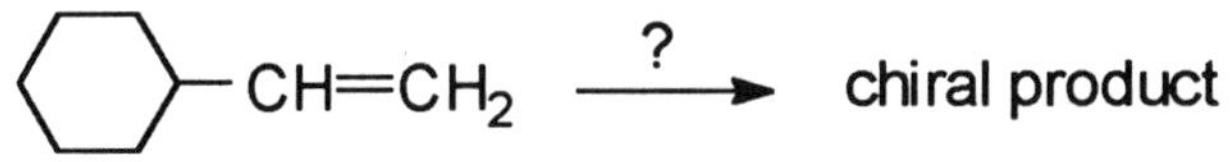

I. $KMnO_4$
II. $Hg(OAc)_2$ H_2O / $NaBH_4$
III. BH_3 then H_2O_2, NaOH
IV. HBr / peroxides

a) I, II b) I, III c) II, IV d) III, IV

21. Calculate the enantiomeric excess for the reaction that gives 75% of the S-form and 25% of the R-form.

a) 25% b) 33 1/3% c) 50% d) 75%

22. Which of the following can be resolved into a pair of enantiomers?

a) cis-1,2-dichlorocyclobutane
b) trans-1,2-dichlorocyclobutane
c) cis-1,3-dichlorocyclobutane
d) trans-1,3-dichlorocyclobutane

23. The specific rotation of dextrarotatory tartaric acid is +12.7 degrees. A mixture of dextrarotatory and levorotatory tartaric acid has a specific rotation of +6.35 degrees. What is the optical purity of the mixture?

a) 25% b) 33 1/3% c) 50% d) 75%

24. A chiral compound, C_5H_8, on catalytic hydrogenation yields an achiral compound, C_5H_{10}. What is the original chiral compound?

a) 1-methylcyclobutene
b) 3-methylcyclobutene
c) 1,2-dimethylcyclopropene
d) cyclopentene

25. A terminal alkyne, C_6H_{10}, loses its chirality when converted to C_6H_{14} by catalytic hydrogenation. What is the alkyne?

a) 1-hexyne
b) 3-methyl-1-pentyne
c) 4-methyl-1-pentyne
d) 3,3-dimethyl-1-butyne

Answers

1. d)
2. b)
3. d)
4. d)
5. c)
6. b)
7. d)
8. c)
9. a)
10. a)
11. c)
12. c)
13. c)
14. a)
15. b)
16. b)
17. a)
18. c)
19. a)
20. a)
21. c)
22. b)
23. c)
24. b)
25. b)

Chapter 9 Alcohols and Thiols

1. What is the correct IUPAC name for the following structure?

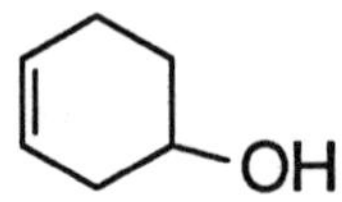

a) cyclohexenol
b) 3-cyclohexen-1-ol
c) 1-cyclohexen-4-ol
d) 4-cyclohexenol

2. What is the correct structure for 5-chloro-E-2-hexen-1-ol?

a) 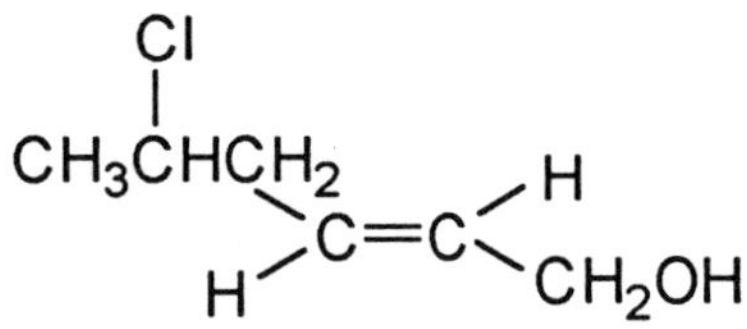

b) 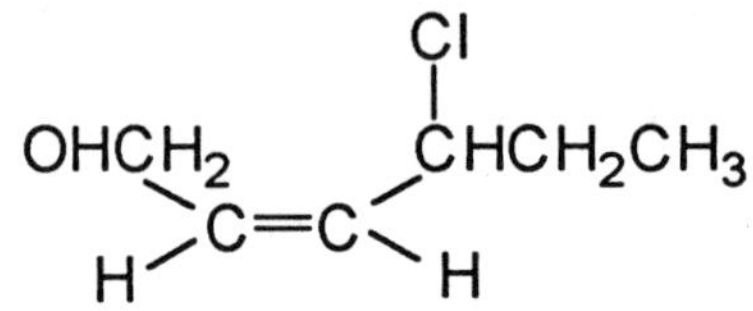

c) 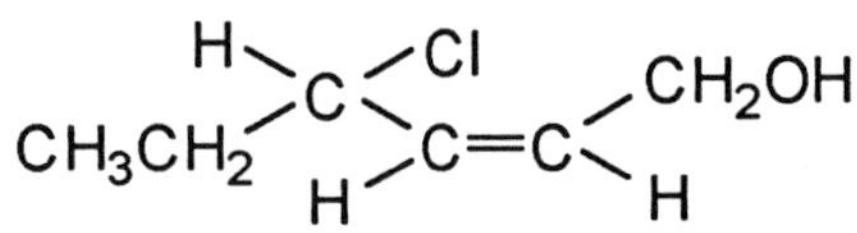

d) 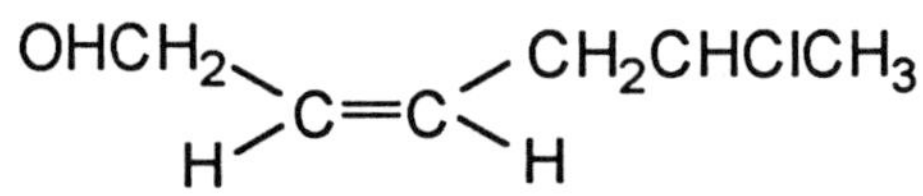

3. Which of the following structures have the correct common name?

$(CH_3)_2CHCH_2OH$	CH_3CH_2SH	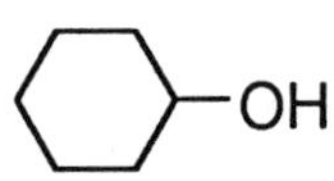	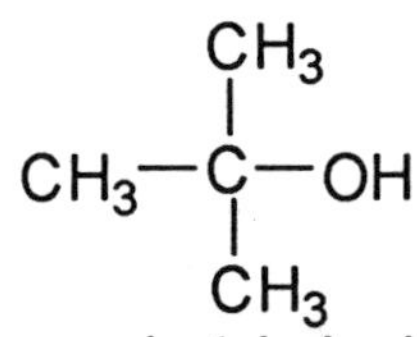
isobutyl alcohol	butyl mercaptan	cyclohexyl alcohol	sec-butyl alcohol
I	II	III	IV

a) I, II
b) I, III
c) III, IV
d) I, IV

4. Which of the following are secondary alcohols?

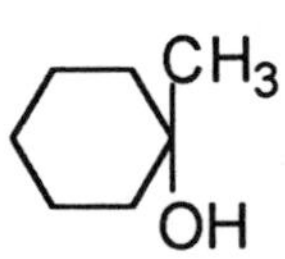	$(CH_3)_3COH$	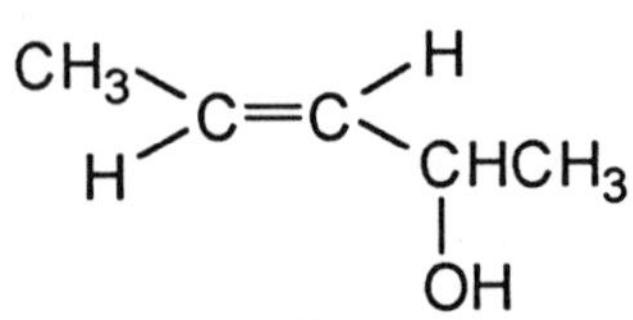	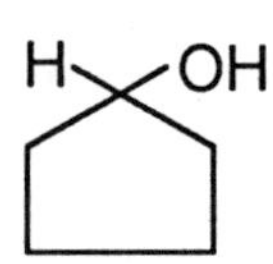
I	II	III	IV

a) I, III
b) II, III
c) III, IV
d) I, IV

5. Arrange the following compounds in the order of increasing polarity (least first).

CH_3CH_2OH	CH_3SH	H_2O	CH_3OCH_3
I	II	III	IV

a) III, IV, II, I b) II, IV, III, I c) II, IV, I, III d) I, III, II, IV

6. Arrange the following compounds in the order of increasing acidity (least first).

CH_3CH_2OH	$CH_3CH(Cl)OH$	$CH_3CH(OH)CH_3$	H_2O
I	II	III	IV

a) III, I, II, IV b) I, III, IV, II c) II, III, IV, I d) II, I, III, IV

7. Arrange the following compounds in the order of increasing acidity (least first).

$(CH_3)_2CHOH$	CH_3CH_2SH	H_2S	H_2O
I	II	III	IV

a) II, I, III. IV b) I, IV, II, III c) III, II, I, IV d) II, I, IV, III

8. Arrange the following compounds in the order of increasing boiling point (lowest first).

CH_3OCH_3	H_2O	CH_3CH_2OH	CH_3CH_2SH
I	II	III	IV

a) II, I, III, IV b) I, IV, II, III c) IV, I, III, II d) I, IV, III, II

9. Arrange the following compounds in the order of increasing solubility in water (lowest first).

CH_3OH	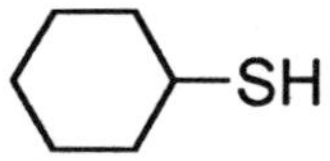	$CH_3CH_2CH_2OH$	$ClCH_2CH_2OH$
I	II	III	IV

a) IV, III, II, I b) II, III, IV, I c) III, IV, II, I d) II, IV, III, I

10. Which of the following equilibria are favored to the right?

I. $CH_3CH_2OH + H_3CC{\equiv}C^{\ominus} \rightleftharpoons CH_3CH_2O^{\ominus} + H_3CC{\equiv}CH$

II. $CH_3CH_2O^{\ominus} + HCl \rightleftharpoons CH_3CH_2OH + Cl^{\ominus}$

III. $CH_3CH_2S^{\ominus} + CH_3CH_2OH \rightleftharpoons CH_3CH_2SH + CH_3CH_2O^{\ominus}$

IV. $NH_3 + CH_3CH_2O^{\ominus} \rightleftharpoons NH_2^{\ominus} + CH_3CH_2OH$

a) I, II   b) II, III   c) III, IV   d) I, III

11. What is the likely mechanism of the following reaction?

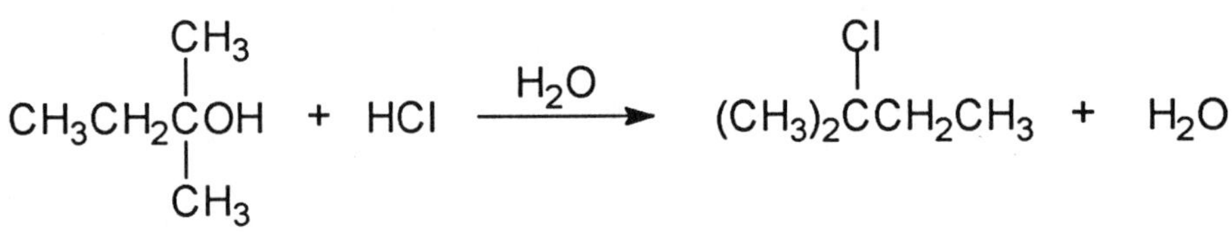

a) SN_1 b) SN_2 c) hydration-dehydration   d) pinacol rearrangement

12. Which of the following alcohols will form the most stable carbocation?

a) $CH_3\overset{\overset{CH_3}{|}}{C}HCH_2OH$ b) $(CH_3)_3OH$ c) 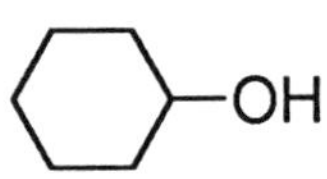   d) CH_3OH

13. How many isomers, including stereoisomers, can be formed from the acid-catalyzed hydration of 4-methyl cyclohexene?

a) 2   b) 4   c) 6   d) 8

14. Which of the following methods for preparing alcohols will NOT give 2-hexanol from 1-hexene?

a) HCl / NaOH
b) oxymercuration-demercuration
c) acid-catalyzed hydration
d) hydroboration / oxidation

15. What are the best conditions for the following alcohol preparation?

a) $\xrightarrow[H_2O]{H_2SO_4}$

b) $\xrightarrow{BH_3}$ $\xrightarrow{H_2O_2\text{, NaOH}}$

c) $\xrightarrow{Hg(OAc)_2\text{, }H_2O}$ $\xrightarrow{NaBH_4}$

d) $\xrightarrow[\text{heat}]{ZnCl_2}$

16. What is the best method for making the following conversion?

$$CH_3CH_2CH_2OH \longrightarrow CH_3\underset{}{\overset{OH}{\overset{|}{C}}}HCH_3$$

a) $\xrightarrow[\text{heat}]{H_3PO_4}$ $\xrightarrow[H_2SO_4]{\text{dilute}}$

b) $\xrightarrow[H_2SO_4\text{ / heat}]{\text{concentrated}}$ $\xrightarrow[\text{NaOH / }H_2O_2]{BH_3}$

c) $\xrightarrow{PCC}$

d) $\xrightarrow{SOCl_2}$ $\xrightarrow{NaOH}$

17. Place the following in the order of increasing strength of the hydrogen bonds (weakest first).

I. HNH-----OH_2

II. H_3N-----HNH_2

III. H_2O-----HOH

IV. H_3N-----HOH

a) I, II, III, IV

b) III, I, IV, II

c) II, IV, I, III

d) I, IV, II, III

18. Which of the following alcohols is most likely to form rearranged products from the reaction with HBr?

a) 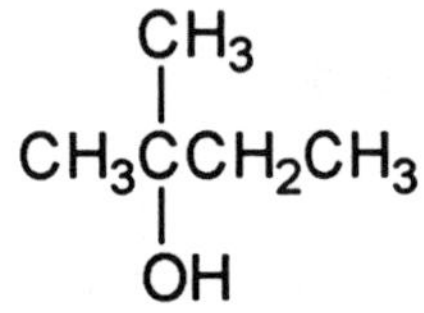

b) 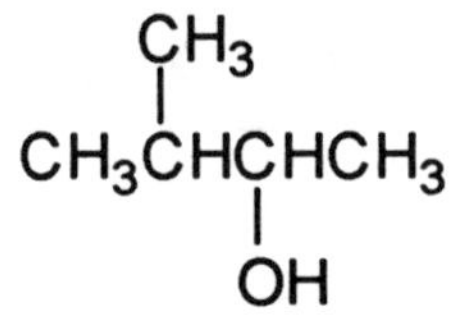

c) 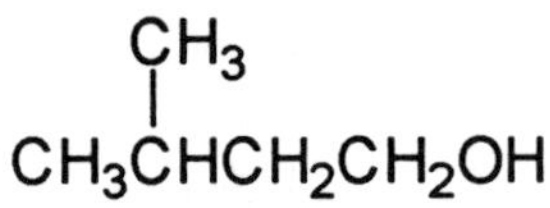

d) $(CH_3)_3CCH_2OH$

19. What are the best conditions for the following conversion?

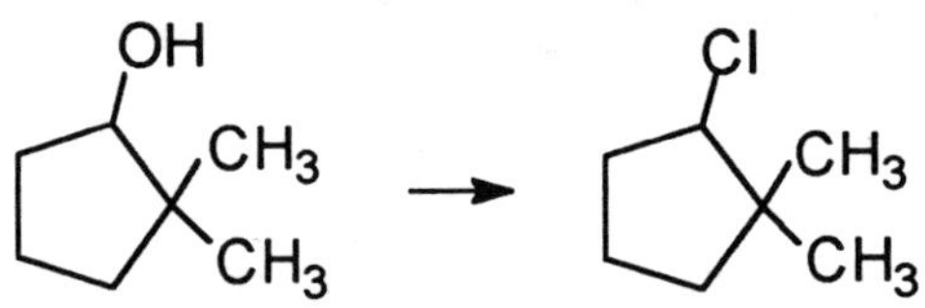

a) $\xrightarrow{SOCl_2}$ b) $\xrightarrow{HCl}$ c) $\xrightarrow{ZnCl_2}$ d) $\xrightarrow{PCl_3}$

20. Which of the following reagents does not effect the following conversion?

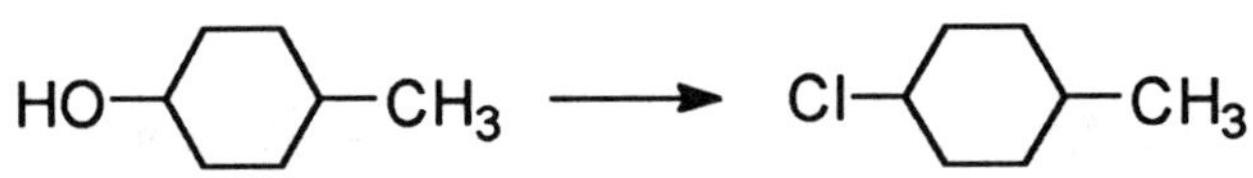

a) $\xrightarrow{HCl}$ b) $\xrightarrow{PCl_3}$ c) $\xrightarrow{SOCl_2}$ d) $\xrightarrow{NaCl}$

21. What is the product from the following reaction?

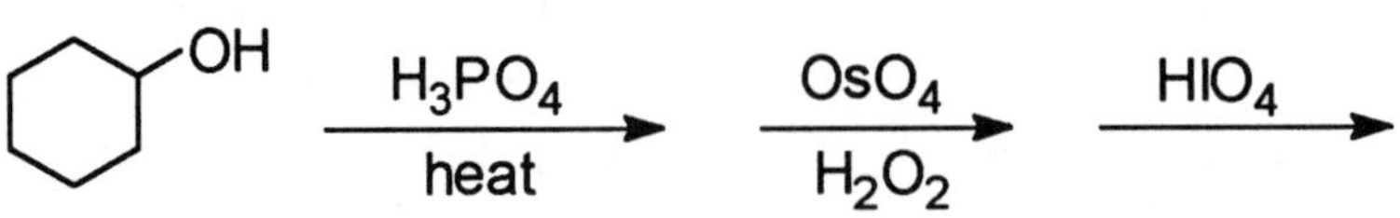

a) 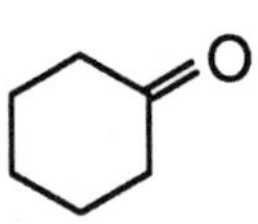 b) 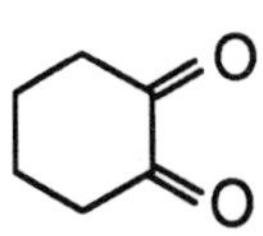 c) 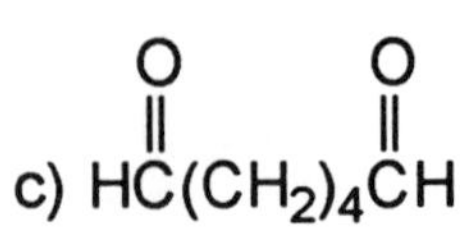 d) $HOOC(CH_2)_4COOH$

22. Which of the following glycols will give an aldehyde with concentrated sulfuric acid and heat?

a) 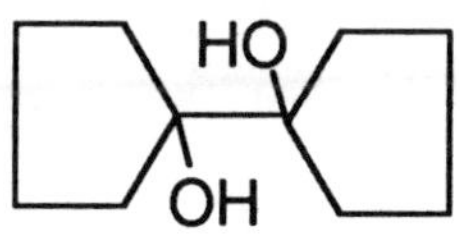

b) 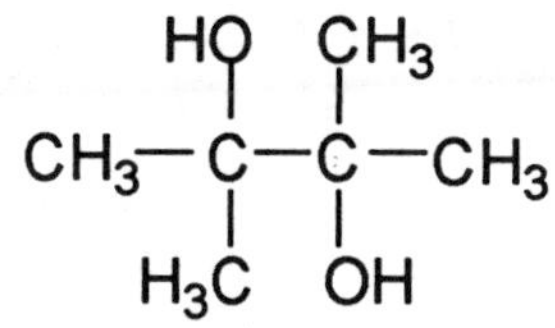

c) 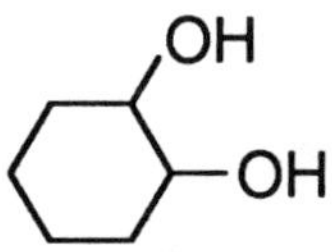

d) 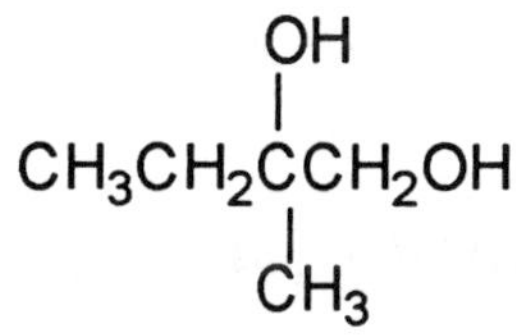

23. What is the product from the following reaction?

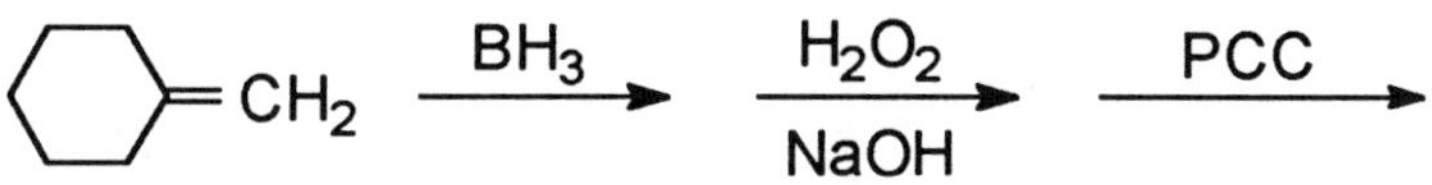

a) 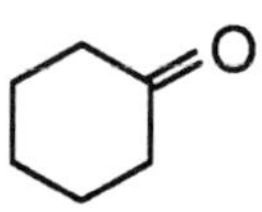

b) 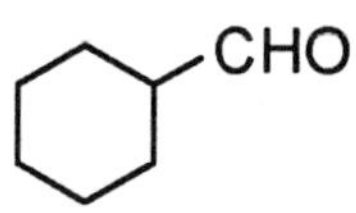

c) 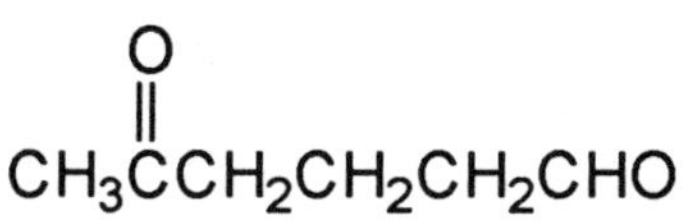

d) 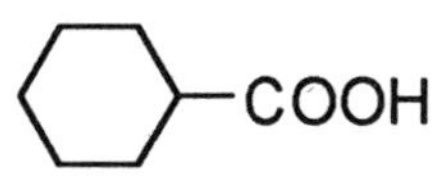

24. Hydroboration of 1-methylcyclohexene followed by treatment with hydrogen peroxide gives product A. Product A is heated with sulfuric acid to give product B. Which of the following structures is the most likely for product B?

a) 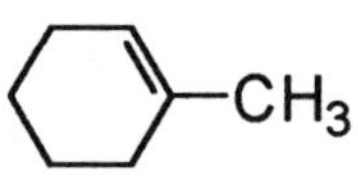

b) 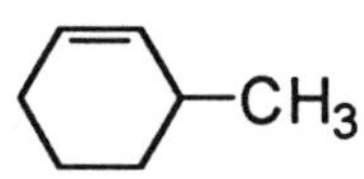

c) 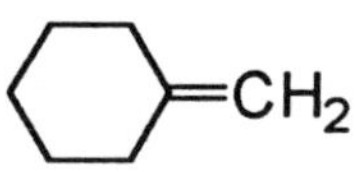

d) 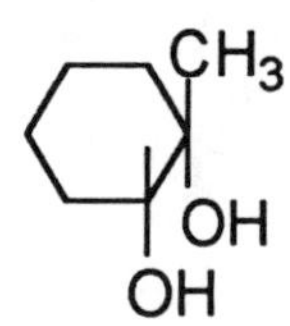

25. Hydroboration of 2-methyl-1-pentene, followed by treatment with hydrogen peroxide and sodium hydroxide, gives the product A. Compound A, when reacted with hydrogen bromide, yields compound B. What is the most likely structure for compound B?

a) $(CH_3)_2\overset{\overset{\large Br}{|}}{C}(CH_2)_2CH_3$

b) $CH_3(CH_2)_2\overset{\overset{\large CH_3}{|}}{C}HCH_2Br$

c) $CH_3CH_2CH{=}C(CH_3)_2$

d) $CH_3(CH_2)_2\overset{\overset{\large Br}{|}}{C}HCH_2OH$

Answers

1. d)
2. a)
3. b)
4. c)
5. c)
6. a)
7. b)
8. d)
9. d)
10. a)
11. a)
12. b)
13. c)
14. d)
15. b)
16. a)
17. a)
18. d)
19. a)
20. d)
21. c)
22. d)
23. b)
24. a)
25. a)

Chapter 10 Alkyl Halides

1. Which of the following structures have the correct IUPAC name?

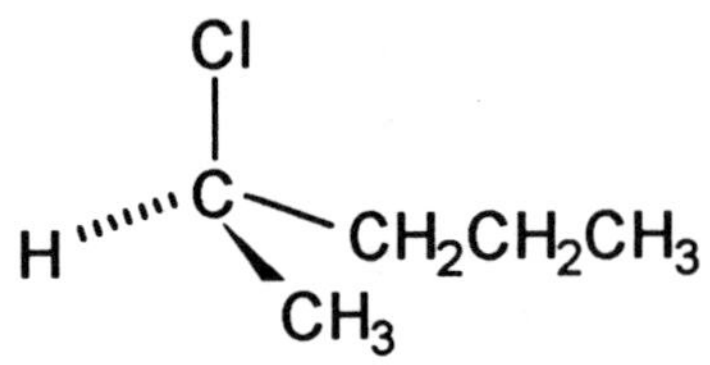

R-2-chloropentane

I

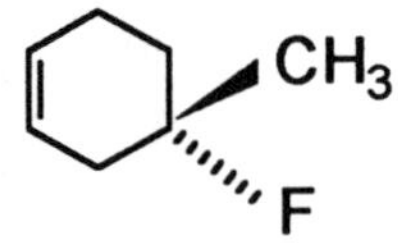

R-4-fluoro-4-methylcyclohexene

II

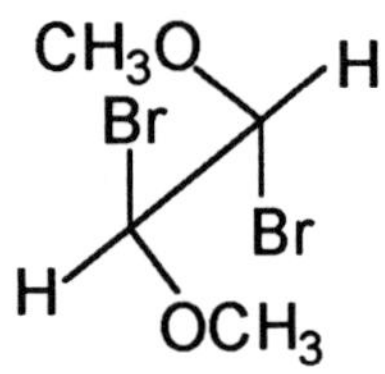

meso-1,2-dibromo-1,2-dimethoxyethane

III

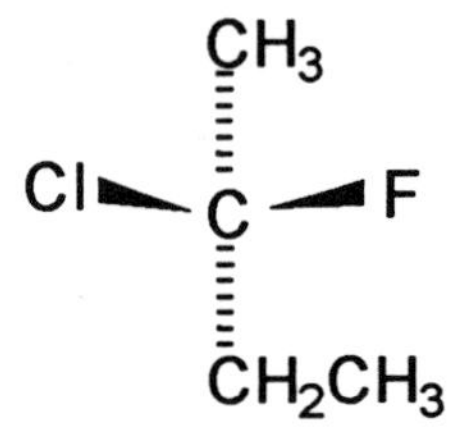

S-2-fluoro-2-chlorobutane

IV

a) I, II b) III, IV c) I, III d) II, IV

2. Which of the following structures have the correct common name?

$CH_2{=}CHCH_2F$

allyl fluoride

I

$(CH_3)_3CCH_2Br$

neopentyl bromide

II

$HCCl_3$

methylene chloride

III

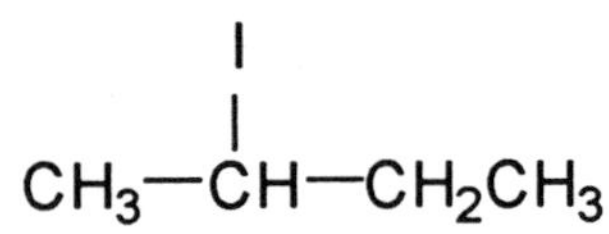

isopropyl iodide

IV

a) I, II b) III, IV c) I, III d) II, IV

3. Which of the following compounds are secondary halides?

isobutyl bromide	2-iodobutane	isopropyl fluoride	neopentyl chloride
I	II	III	IV

a) I, II b) III, IV c) II, III d) I, IV

4. Arrange the following compounds in the order of increasing polarity (least first).

chloroform	acetone	water	carbon tetrachloride
I	II	III	IV

a) I, IV, II, III b) IV, I, II, III c) II, I, IV, III d) II, IV, I, III

5. Which of the following halides is prepared from the following reaction?

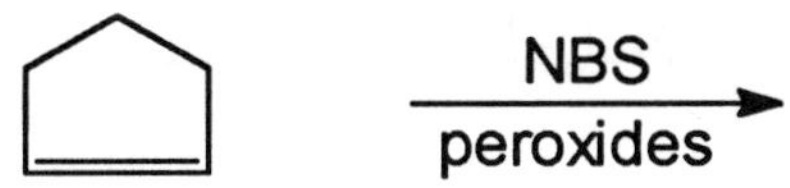

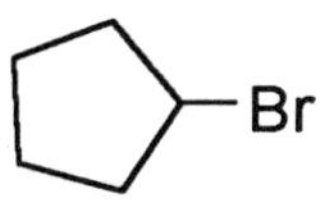

a)

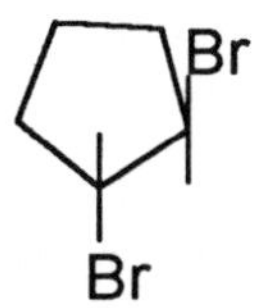

b)

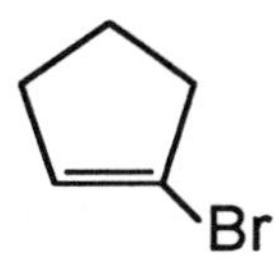

c)

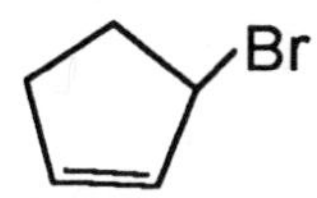

d)

6. What are the best conditions for preparing 2-iodohexane from 1-hexene?

a) I_2 and heat b) HI c) HI and peroxides d) HIO_4

7. What is the order of increasing reactivity for the following nucleophiles (least first)?

CH_3OH	H_2O	$OH^{\ominus}$	$CH_3O^{\ominus}$
I	II	III	IV

a) I, IV, II, III b) IV, II, I, II c) II, I, III, IV d) IV, III, I, II

8. What is the order of increasing reactivity for the following halides in an SN_2 reaction with KI in acetone (least first)?

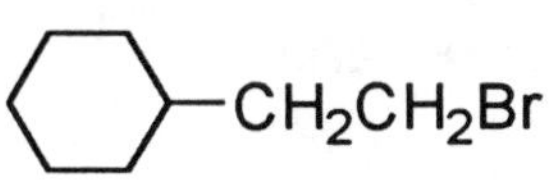

I

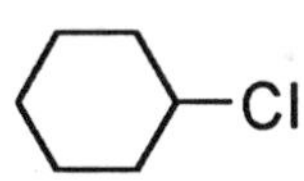

II

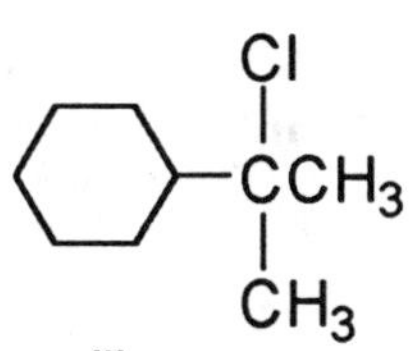

III

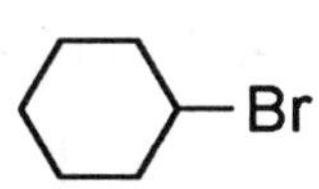

IV

a) I, IV, III, II b) IV, I, III, II c) II, III, I, IV d) III, II, IV, I

9. Which of the following are optimum conditions for an SN_2 reaction?

I. 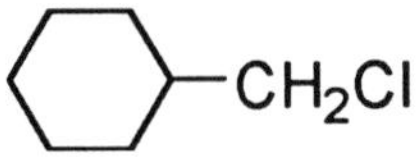 + NaSH $\xrightarrow{\text{acetone}}$

II. 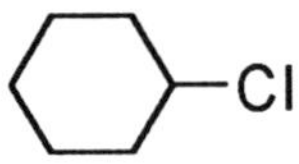 + CH_3OH $\xrightarrow{H_2O}$

III. 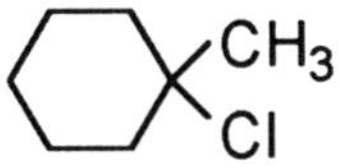 + NaCN $\xrightarrow{\text{DMSO}}$

IV. 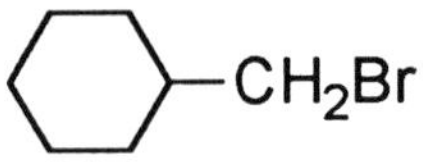 + NaI $\xrightarrow{\text{DMSO}}$

a) I, II b) II, III c) III, IV d) I, IV

10. Which of the following nucleophilic substitution reactions will proceed?

I. NaCl + CH_3CH_2I $\xrightarrow{\text{acetone}}$

II. NH_3 + 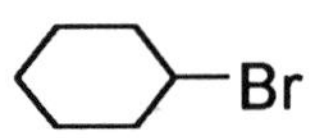 $\xrightarrow{\text{ethanol}}$

III. $CH_3CH_2O^{\ominus}$ + $CH_3CH_2CH_2Cl$ $\xrightarrow{\text{ethanol}}$

IV. LiCl + CH_3I $\xrightarrow{\text{acetone}}$

a) I, II b) II, III c) III, IV d) I, IV

11. Which of the following reactions will proceed with inversion of configuration?

I. R-2-bromohexane $\xrightarrow[CH_3CH_2OH]{^{\ominus}OH}$

II. S-2-iodobutane $\xrightarrow{NaSCH_3}$

III. R-2-bromo-2-ethylpentane $\xrightarrow[CH_3CH_2OH]{^{\ominus}OH}$

IV. R-2-bromohexane $\xrightarrow[\text{heat}]{CH_3OH}$

a) I, II b) III, IV c) II, III d) I, IV

12. What is the product from the following reaction?

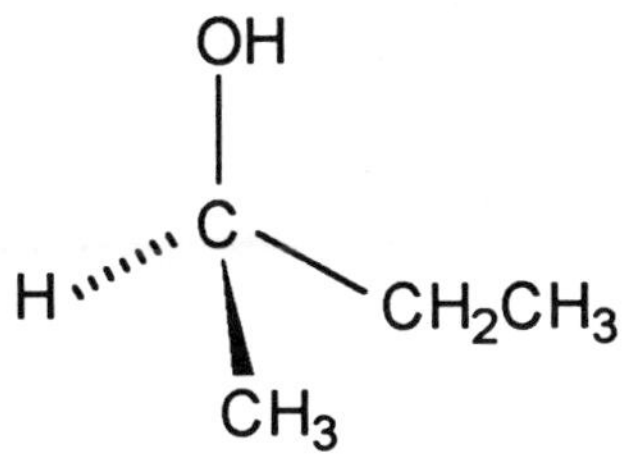

$+ \ SOCl_2 \xrightarrow{\text{diethyl ether}}$

a) R-2-chlorobutane b) S-2-chlorobutane
c) 1-chlorobutane d) mixture of a) and b)

13. Place the following species in the order of increasing order of reactivity as leaving groups (least first).

$^{\oplus}NH_3$	$^{\ominus}OH$	$^{\oplus}OH_2$	$CH_3C(=O)O^{\ominus}$
I	II	III	IV

a) I, III, IV, II b) IV, II, III, I c) III, I, IV, II d) II, IV, I, III

14. What is the order of increasing rate of solvolysis for the following halides (slowest first).

$(CH_3)_3CCl$	$(CH_3CH_2)_2CHCl$	$(CH_3)_2CHCl$	$CH_3CH_2CH_2Cl$
I	II	III	IV

a) IV, III, II, I b) I, II, III, IV c) III, II, I, IV d) II, III, I, IV

15. Which of the following conditions is optimum for an SN_1 reaction?

a) $CH_3CH_2Cl + NaCN \xrightarrow{\text{acetone}}$

b) $(CH_3)_2CHCl + CH_3OH \xrightarrow{H_2O}$

c) $CH_3CH_2CH_2Br + NaSH \xrightarrow{\text{DMSO}}$

d) $(CH_3)_3CCl + NaCN \xrightarrow{\text{DMSO}}$

16. In the following reaction, if the concentration of t-butyl bromide and ethanol are doubled, what effect does this have on the rate of reaction?

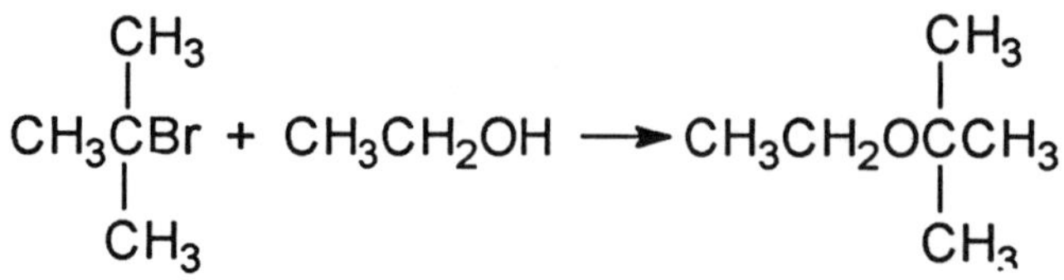

a) no change b) doubles c) triples d) quadruples

17. In the following reaction, if the concentration of hexyl bromide and $NaSCH_3$ are doubled, what effect does this have on the rate of reaction?

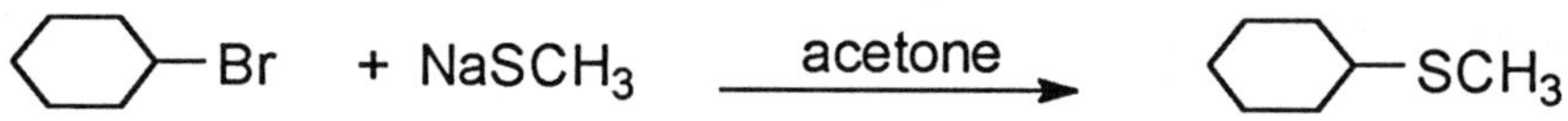

a) no change b) doubles c) triples d) quadruples

18. Which of the following statements characterize the reaction of R-3-chlorocyclopentene with sodium iodide in acetone?

I. The reaction involves a carbocation intermediate.
II. The reaction involves inversion of configuration.
III. The reaction involves retention of configuration.
IV. The reaction gives predominately a racemic product.

a) I b) II c) I, IV d) I, III

19. Which of the following statements apply to an SN_1 reaction?

I. The reaction is first order in alkyl halide and first order in the nucleophile.
II. The order of reactivity is methyl > 1°> 2° > 3°.
III. The reaction is first order in alkyl halide and zero order in the nucleophile.
IV. Rearrangements are common.

a) I, II b) III, IV c) I, IV d) III

20. Which of the following statements apply to an SN_2 reaction?

I. The reaction is first order in alkyl halide and first order in the nucleophile.
II. The order of reactivity is methyl > 1°> 2° > 3°.
III. The reaction is first order in alkyl halide and zero order in the nucleophile.
IV. Rearrangements are common.

a) I, II b) III, IV c) I, IV d) II, IV

21. Which of the following are optimum conditions for an E_2 reaction?

a) $CH_3I + F^{\ominus} \xrightarrow{DMSO}$

b) 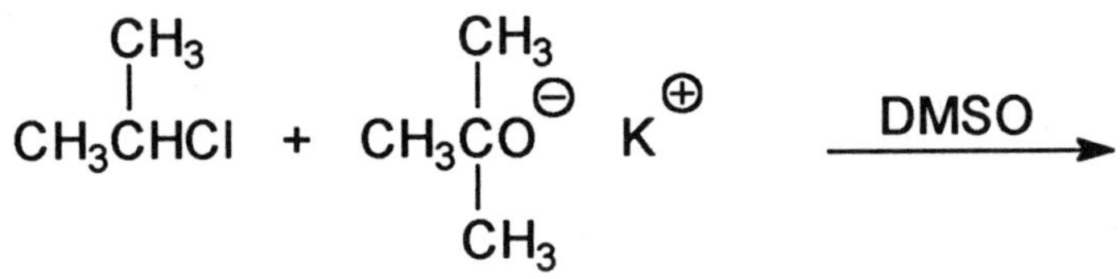

c) 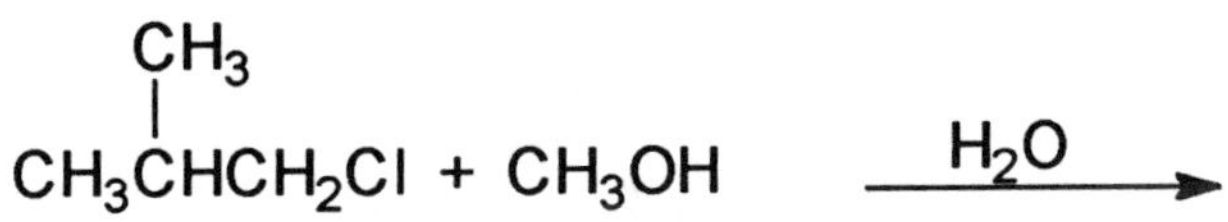

d) 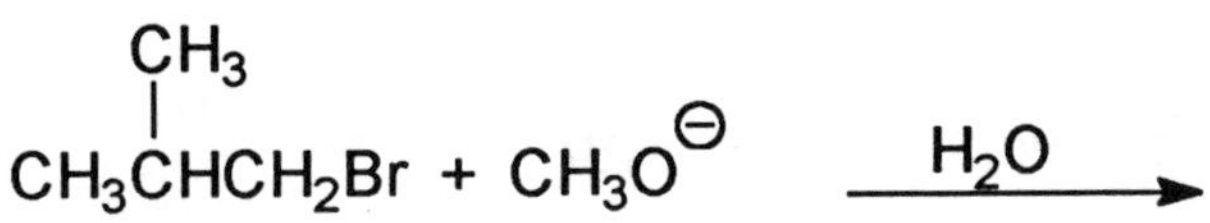

22. What is the principal product from the Zaitsev elimination of 2-chloro-3-phenylbutane?

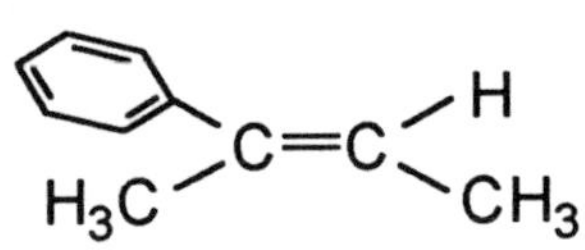

a)

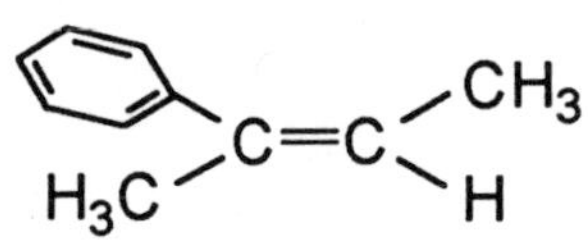

b)

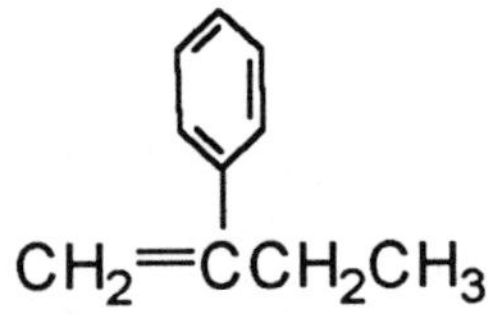

c)

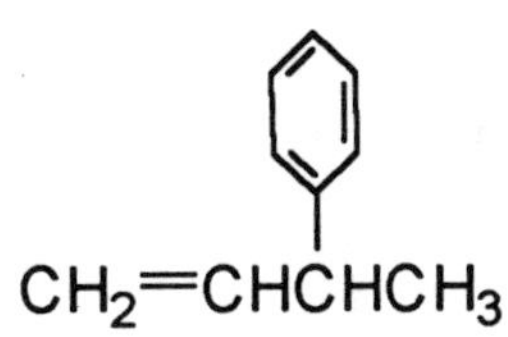

d)

23. What is the principal product from an E_2 reaction of 3-bromo-3,4-dimethylhexane?

a) 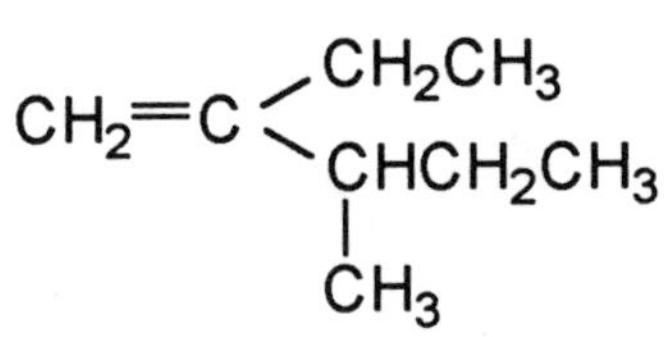

b) 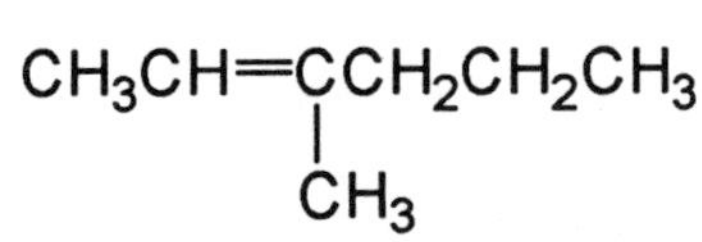

c) 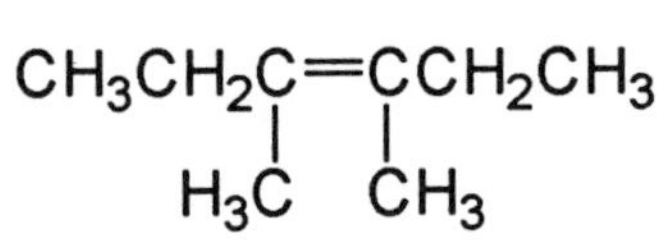

d) 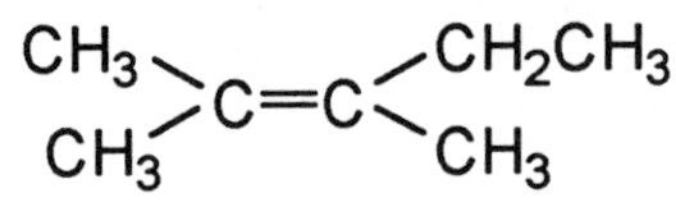

24. Which of the following statements are true for an E_1 reaction?

I. The reaction is first order in alkyl halide and zero order in base.
II. The reaction is first order in alkyl halide and first order in base.
III. Rearrangements are common.
IV. The order of reactivity is $1^\circ > 2^\circ > 3^\circ$.

a) I,III b) II c) I, III, IV d) II, IV

25. Which of the following conditions is best for preparing isopropyl methyl ether?

a) $Na^{\oplus}\ {}^{\ominus}OCH_3 + (CH_3)_2CHBr \longrightarrow$

b) $Na^{\oplus}\ {}^{\ominus}OCH(CH_3)_2 + CH_3Br \longrightarrow$

c) $CH_3OH + (CH_3)_2CHBr \xrightarrow{\text{heat}}$

d) $CH_3OH + (CH_3)_2CHOH \xrightarrow[\text{catalyst}]{\text{acid}}$

Answers

1. c)
2. a)
3. c)
4. b)
5. d)
6. b)
7. c)
8. d)
9. d)
10. b)
11. a)
12. a)
13. d)
14. a)
15. b)
16. b)
17. d)
18. b)
19. b)
20. a)
21. b)
22. b)
23. c)
24. a)
25. b)

Chapter 11 Ethers and Epoxides

1. Which of the following structures have the correct name?

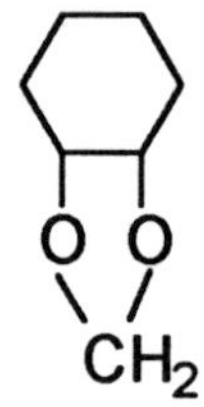

cis-1,2-dimethoxycyclohexane

I

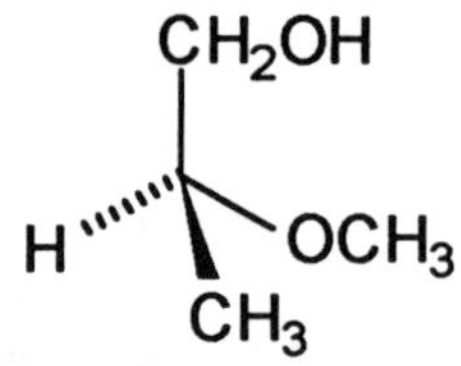

R-2-methoxy-1-propanol

II

$(CH_3)_2CHOCH_3$

isopropyl methyl ether

III

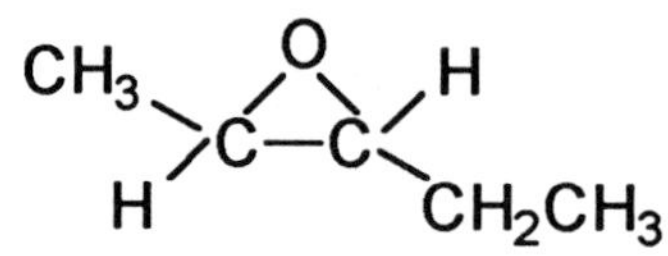

trans-2-ethyl-3-methyloxirane

IV

a) I, II b) III, IV c) II, III d) I, IV

2. Which of the following structures have the correct name?

$CH_3OCH_2CH_2OCH_3$

THF

I

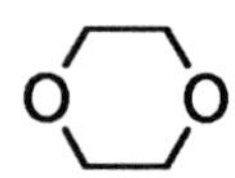

1,4-dioxane

II

DME

III

$(CH_3CH_2)_2O$

diethyl ether

IV

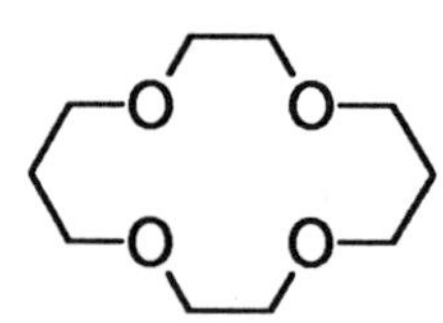

18-crown-6

V

a) I, III b) II, IV c) II, V d) IV, V

3. Place the following compounds in the order of increasing solubility in water (least first).

$CH_3(CH_2)_3OCH_3$	CH_3OCH_3	$CH_3OCH_2CH_2OCH_3$	$CH_3(CH_2)_4OH$
I	II	III	IV

a) I, III, II, IV   b) III, I, IV, II   c) I, IV, II, III   d) IV, I, III, II

4. Place the following compounds in the order of increasing base strength (least first).

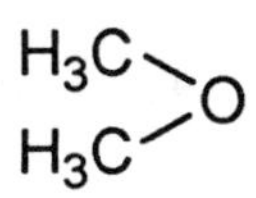
I

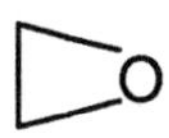
II

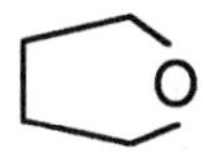
III

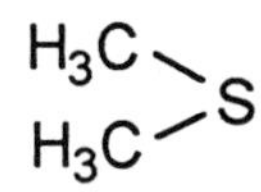
IV

a) II, III, I, IV   b) I, III, II, IV   c) IV, III, I, II   d) IV, I, III, II

5. What are the best conditions for preparing tert-butyl cyclohexyl ether?

a) 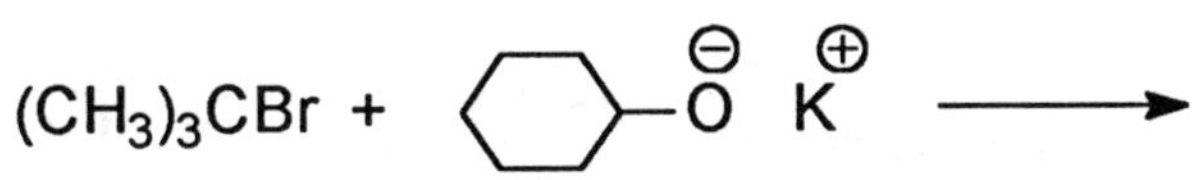

b) 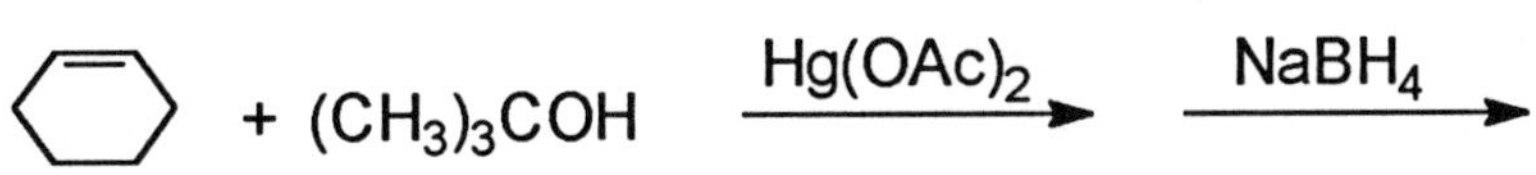

c) 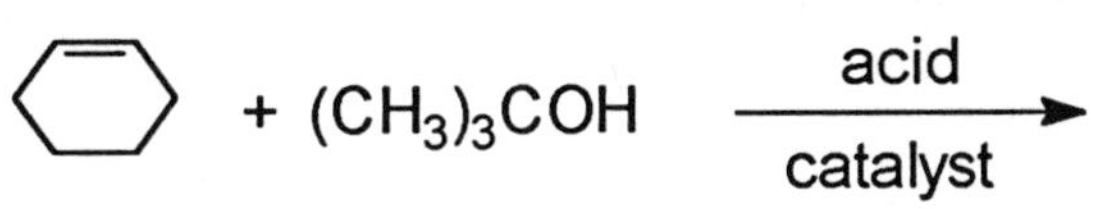

d) 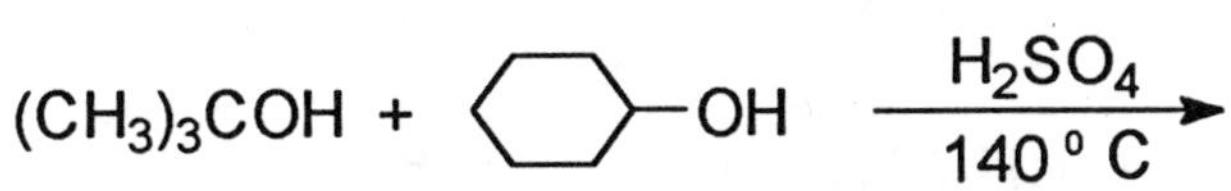

6. Which of the following conditions are NOT good for preparing dibutyl ether?

I. $CH_3(CH_2)_3Br + CH_3(CH_2)_3O^{\ominus}\ K^{\oplus} \longrightarrow$

II. $CH_3CH_2CH{=}CH_2 + CH_3(CH_2)_3OH \xrightarrow{Hg(OAc)_2} \xrightarrow{NaBH_4}$

III. $CH_3CH_2CH{=}CH_2 + CH_3(CH_2)_3OH \xrightarrow[\text{catalyst}]{\text{acid}}$

IV. $CH_3(CH_2)_3OH \xrightarrow[140^{\circ}C]{H_2SO_4}$

a) I, IV b) II, III c) I, II d) III, IV

7. What are the best conditions for preparing R-2-ethoxypentane?

a) $CH_3CH_2CH_2C(CH_3){=}CH_2 \xrightarrow[\text{catalyst}]{\text{acid}}$

b) $CH_3CH_2CH_2CH{=}CH_2 + CH_3CH_2OH \xrightarrow{Hg(OAc)_2} \xrightarrow{NaBH_4}$

c) 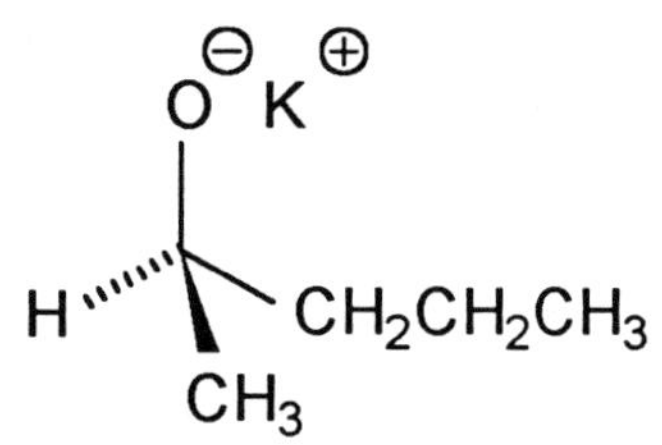 $+ CH_3CH_2Br \longrightarrow$

d) 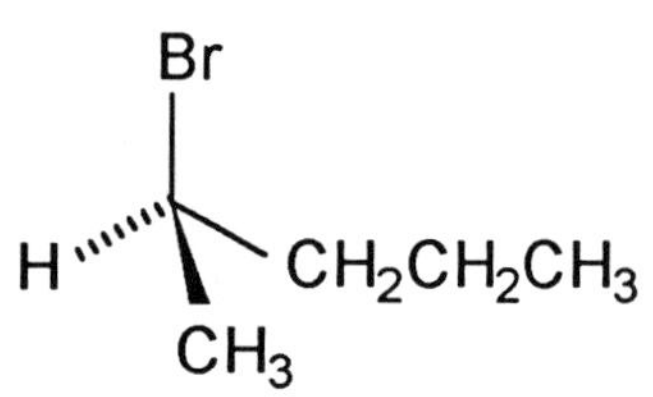 $+ CH_3CH_2O^{\ominus}\ K^{\oplus} \longrightarrow$

8. What are the best conditions for preparing tetrahydrothiophene?

a) 2 CH_3CH_2Cl + Na_2S $\longrightarrow$

b)  + Na_2S $\longrightarrow$

c) CH_3CH_2S Na + CH_3CH_2I $\longrightarrow$

d) $ClCH_2CH_2CH_2CH_2Cl$ + Na_2S $\longrightarrow$

9. What are the best conditions for preparing trans-2-butene oxide?

a) E-2-butene + 3-chloroperoxybenzoic acid
b) Z-2-butene + 3-chloroperoxybenzoic acid
c) 1-butene + chlorine and water followed by aqueous sodium hydroxide
d) 1,4-dichlorobutane + sodium sulfide

10. What are the products from the following reaction?

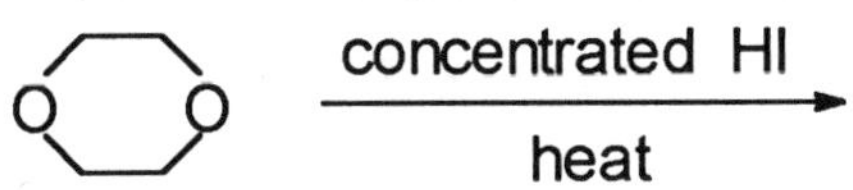

a) 2 CH_3CH_2I

b) 2 ICH_2CH_2I

c) ICH_2CH_2OH + ICH_2CH_2I

d) CH_3CH_2OH + CH_3CH_2I

11. What are the products from the SN_2 reaction of ethyl methyl ether with HBr?

a) CH_3CH_2Br + CH_3Br

b) CH_3CH_2OH + CH_3Br

c) $CH_2{=}CH_2$ + CH_3Br

d) CH_3CH_2Br + CH_3OH

12. What is the major product from the acid hydrolysis of cis-1,2 epoxy-cyclohexene?

a) 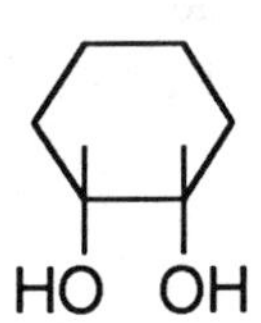

b) 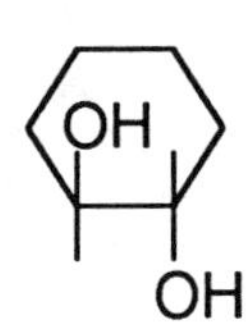

c) 

d) 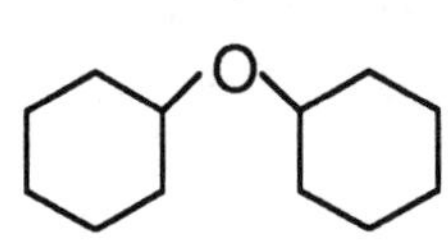

13. Which of the following ethers give the product shown in the reaction?

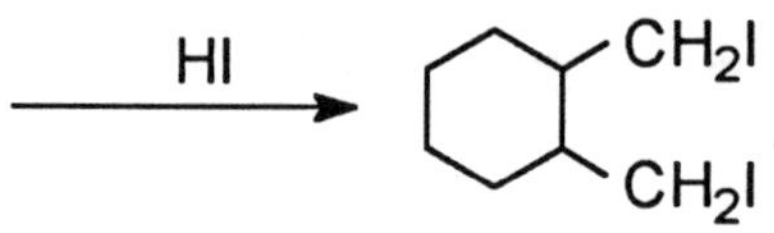

a) 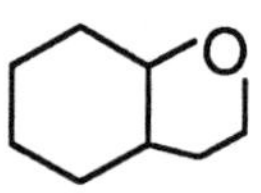 b) 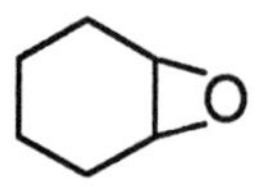 c) 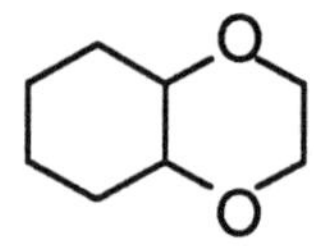 d) 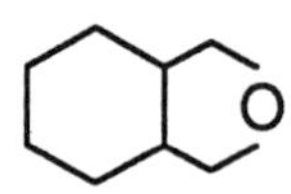

14. What are the best conditions for the following reaction?

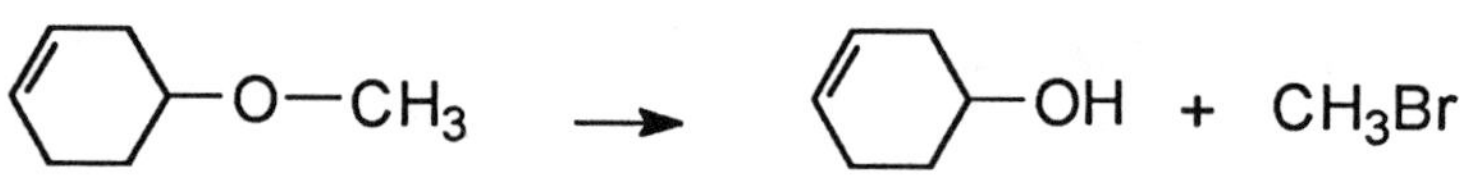

a) HBr b) BBr_3 followed by H_2O c) H_2SO_4 d) HI

15. What are the products from the reaction of diisopropyl ether with BBr_3 followed by H_2O?

a) $CH_3CH_2CH_2Br$

b) $CH_2{=}CHCH_3$

c) 2 $CH_3CH(CH_3)Br$

d) $CH_3CH(CH_3)Br$ + $CH_3CH(CH_3)OH$

16. What are the best conditions for the following conversion?

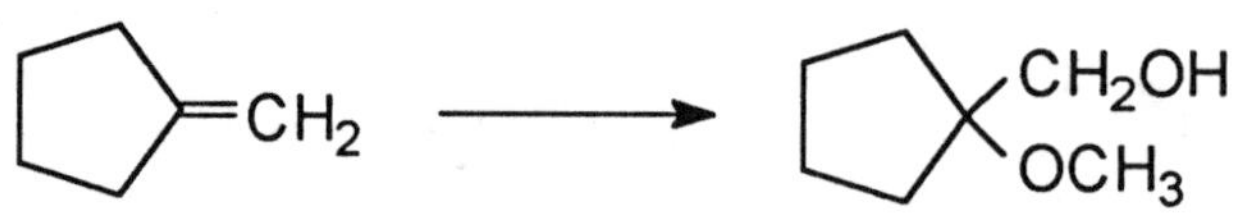

a) chlorine and water, followed by aqueous sodium hydroxide
b) perbenzoic acid, followed by methanol and acid
c) perbenzoic acid, followed by methanol and sodium methoxide
d) chlorine and water, followed by sodium hydroxide, then methanol and acid

17. What is the principal product from the following reaction?

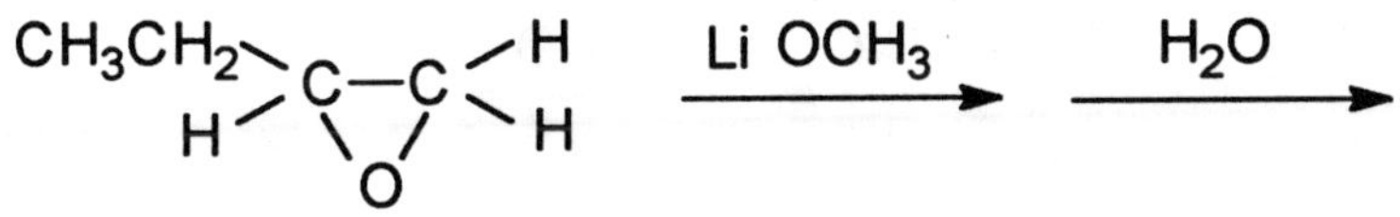

a) 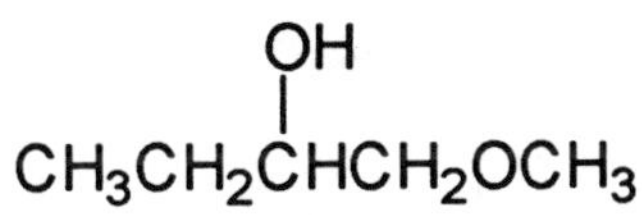
b) 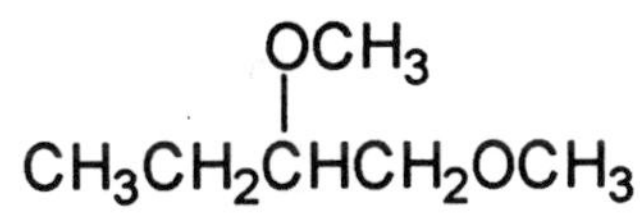
c) 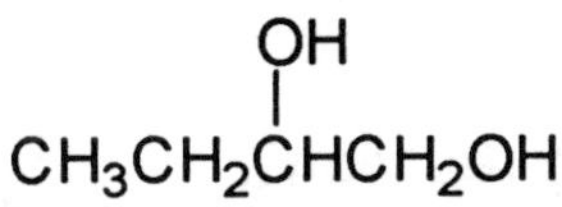
d) 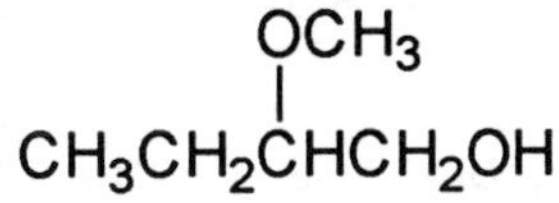

18. What is the principal product from the following series of reactions?

$$CH_3CH_2I \xrightarrow{Na_2S} \xrightarrow{\text{excess } H_2O_2}$$

a) 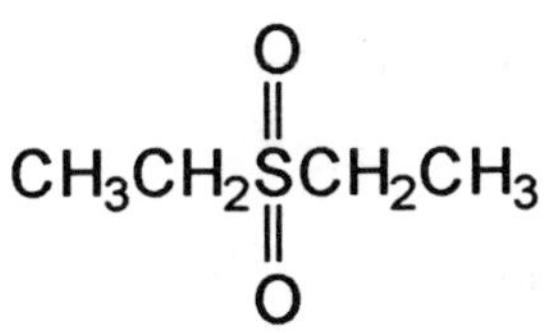
b) 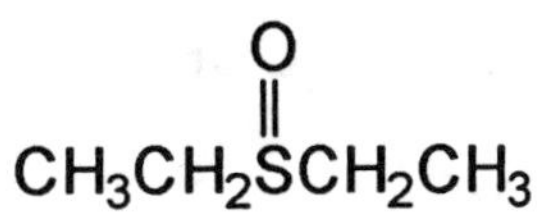
c) 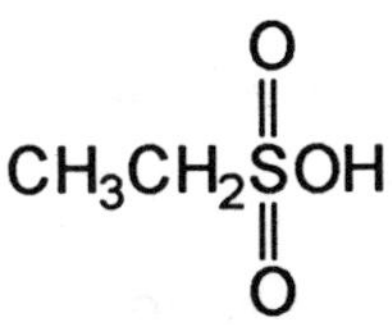
d) 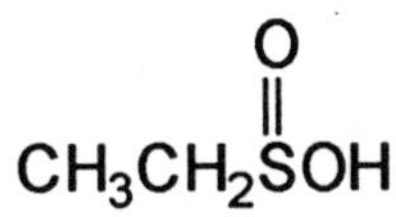

19. What is the mechanism that best describes the following reaction?

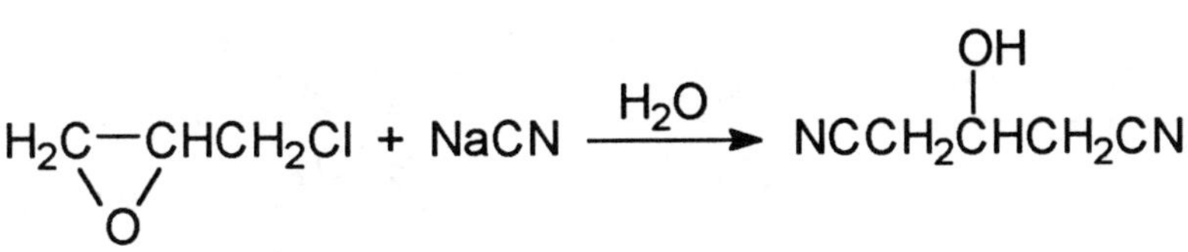

a) SN_1 b) SN_2 c) E_1 d) E_2

20. What is the major product from the following reaction?

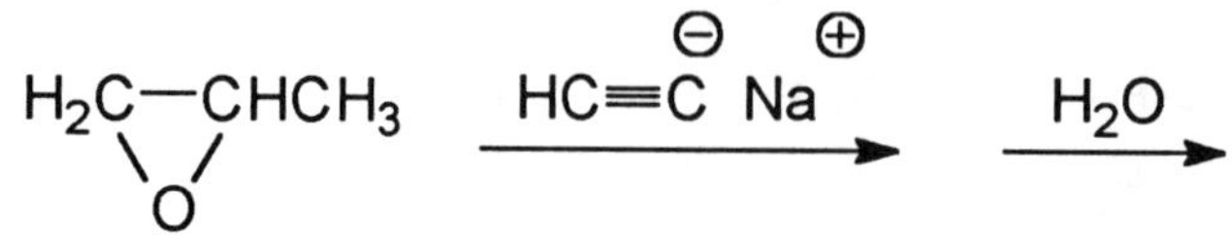

a) $HC{\equiv}CCH_2CH(OH)CH_3$ b) $HC{\equiv}CCH(OH)CH_2CH_3$

c) $HC{\equiv}CCH_2CH_2CH_2OH$ d) $CH_3OC{\equiv}CCH_3$

21. What is the principal product from the following reaction?

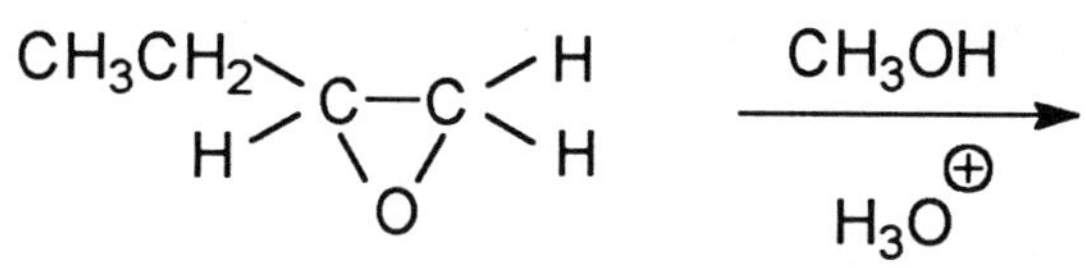

a) $CH_3CH_2CH(OH)CH_2OCH_3$ b) $CH_3CH_2CH(OCH_3)CH_2OCH_3$

c) $CH_3CH_2CH(OH)CH_2OH$ d) $CH_3CH_2CH(OCH_3)CH_2OH$

22. What is the principal product from the reaction of trans-1,2-diethyloxirane with sodium hydroxide?

a) 3R,4R-3,4-hexanediol b) 3S,4S-3,4-hexanediol
c) 3R,4S-3,4-hexanediol d) a mixture of a) and b)

23. What is the principal product from the following series of reactions starting with cyclohexene?

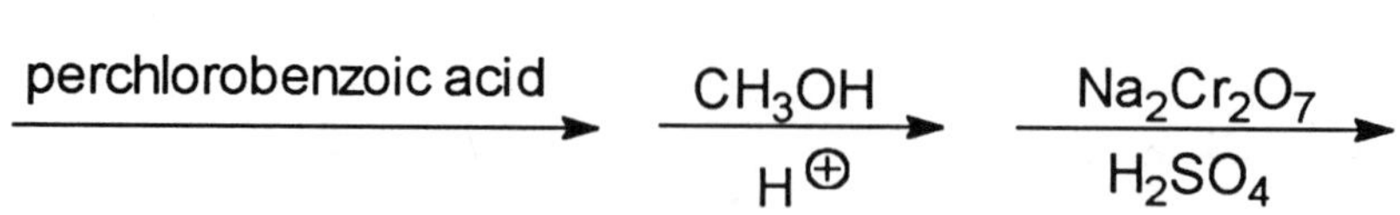

a) 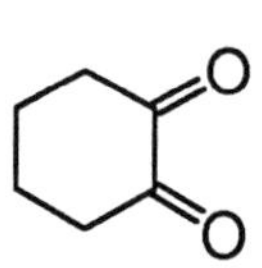
b) 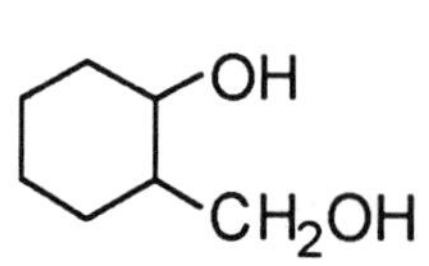

c) 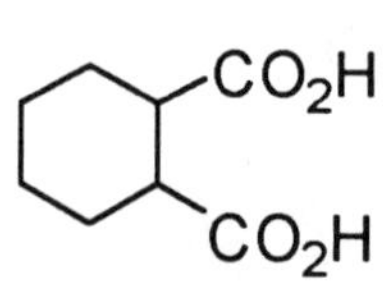

d) 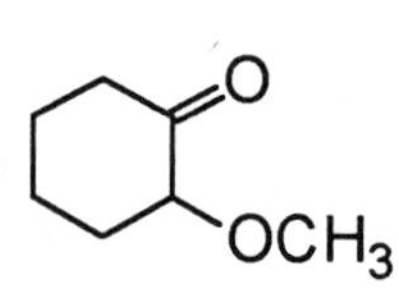

24. What is the principal product from the following reaction?

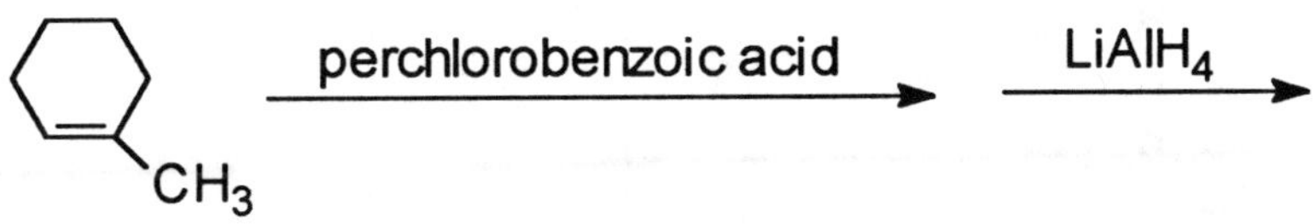

a) 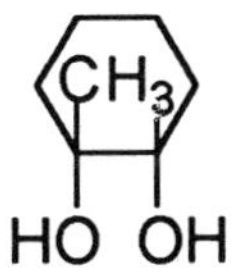
b) 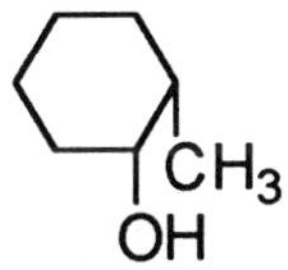
c) 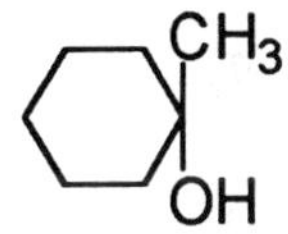
d) 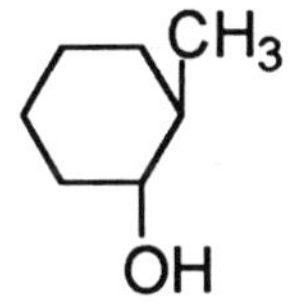

25. What is the principal product from the following reaction?

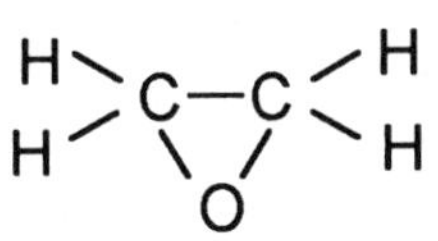 + $NH_2CH_2CH_2NH_2$ ⟶

a) $HOCH_2CH_2NH_2$

b) 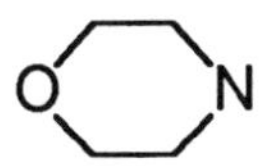

c) $HOCH_2CH_2NHCH_2CH_2OH$

d) $HOCH_2CH_2NHCH_2CH_2NHCH_2CH_2OH$

Answers

1. b)
2. b)
3. c)
4. d)
5. b)
6. b)
7. a)
8. d)
9. a)
10. a)
11. a)
12. b)
13. d)
14. b)
15. d)
16. b)
17. a)
18. a)
19. b)
20. a)
21. d)
22. c)
23. d)
24. c)
25. d)

Chapter 12 Mass Spectrometry

1. A mass spectrum of compound X shows one peak at m/e = 78 and another peak at m/e = 80 with 1/3 the intensity of the former. What is the most likely structure for compound X?

a) $BrCH_2CH_2CH_2Cl$

b) $CH_3CH_2CH_2Cl$

c) 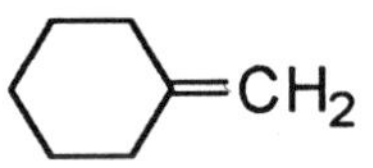

d) 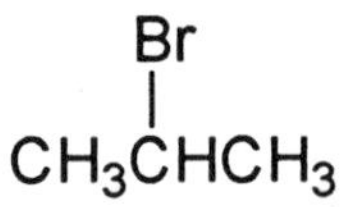

2. Which of the following compounds give a molecular ion having an odd number value for m/e?

CH_3CH_2Br	$CH_3CH_2NH_2$	$H_2NCH_2NH_2$	CH_2Br	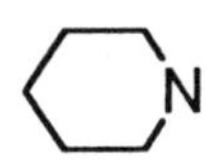	CH_3CH_2OH
I	II	III	IV	V	VI

a) I, IV, VI b) II, V c) I, IV d) II, III, V

3. The mass spectrum of a compound showed a molecular ion peak at M and another at M+2 in the ratio of approximately 1:1. The compound has which of the following heteroatoms?

a) O b) Cl c) Br d) N

4. The mass spectrum of a compound showed a molecular ion peak at M and another at M+2 in the ratio of 3:1. The compound has which of the following heteroatoms?

a) O b) Cl c) Br d) N

5. Which of the following statements best describes the base peak in a mass spectrum?

a) The peak from the most stable fragmentation ion.
b) The peak from the species that has the isotope with the highest atomic no.
c) The peak of highest intensity.
d) The peak from the molecule minus an electron.

6. The base peak for a hydrocarbon is 57. Which of the following is the most likely compound?

a) 2,2,4-trimethylpentane b) octane
c) methylcyclopentane d) 2-methylpentane

7. What kinds of compounds give a strong peak at M - 18?

a) halides b) alcohols
c) cycloalkanes d) cyclohexenes

8. The mass spectrum for a hydrocarbon shows a series of peaks separated by 14 m/e units. Which of the following best describes the hydrocarbon?

a) a branched chain b) a cycloalkane
c) a straight chain d) an alkene

9. Which of the following compounds probably undergo a McLafferty rearrangement?

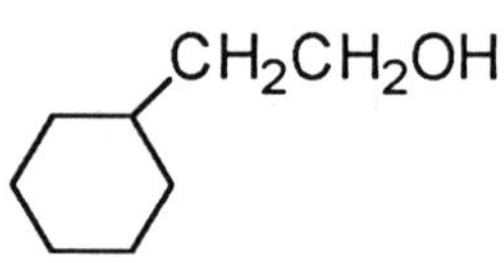
I

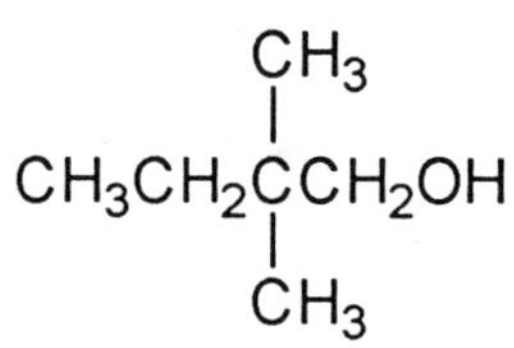
II

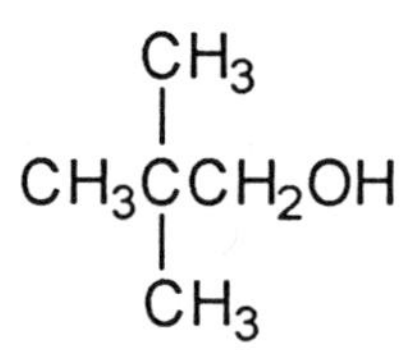
III

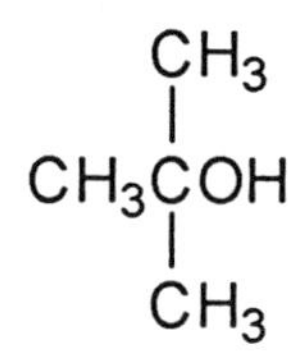
IV

a) I, II b) III, IV c) I, IV d) II, III

10. What kinds of compounds give an allylic cation as a common fragmentation ion?

a) alkenes b) alkynes c) cycloalkanes d) cyclohexenes

11. The base peak for a series of compounds is 41. Which of the following compounds is most likely in this group?

a) 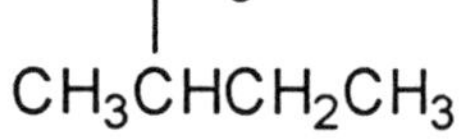

b) 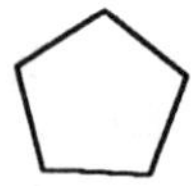

c) $CH_3CH_2CH_2CH_3$

d) $CH_2{=}CHCH_2CH_2CH_3$

12. What is the radical cation formed from 1,3-butadiene?

a) $CH_2{=}CH{-}\overset{\oplus}{C}H{-}\overset{\bullet}{C}H_2$ b) $CH_2{-}\overset{\bullet}{C}H{=}CH{-}\overset{\oplus}{C}H_2$

c) $CH_2{-}\overset{\oplus}{C}H{=}CH{-}\overset{\bullet}{C}H_2$ d) $CH_2{-}CH{=}CH{-}\overset{\bullet\oplus}{C}H_2$

13. What is a likely fragmentation ion from propyne?

a) $CH_2{=}CH{-}\overset{\oplus}{C}H_2$ b) $CH_2{=}C{=}\overset{\oplus}{C}H$

c) $CH_3{-}C{\equiv}\overset{\oplus}{C}$ d) $CH{\equiv}C{-}\overset{\oplus}{C}H_2$

14. The mass spectrum of compound A shows two peaks of nearly equal intensity at m/e = 122 and 124, and a very small peak at 123. Which of the following is most likely compound A?

a) $BrCH_2CH_2CH_2Br$ b) $CH_3CH_2CH_2Cl$

c) $CH_3CH_2CH_2OH$ d) $CH_3\underset{\substack{|\\ Br}}{C}HCH_3$

15. What is the base peak expected for methylcyclopentane?

a) 15 b) 28 c) 56 d) 84

16. What is a characteristic fragmentation peak for 2,2,4-trimethylpentane?

a) 28 b) 43 c) 57 d) 84

17. Which peak in the mass spectra could distinguish between 1-butanol and 2-butanol?

a) 74 b) 28 c) M - 18 d) M + 2

18. Which of the following compounds gives a base peak at 43?

a) $CH_3(CH_2)_4CH_3$ b) 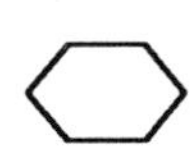

c) $(CH_3)_2CHCH_2CH_2CH_3$ d) $(CH_3)_2CHCH(CH_3)_2$

19. Which of the following compounds has the greatest intensity for the ratio of the peaks, (m+2)/m?

a) 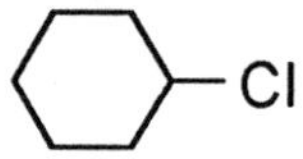 b) 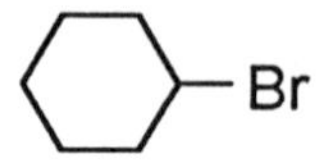 c) 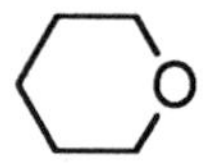 d) 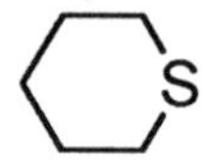

20. An alkyl chloride can be detected by which of the following in a mass spectrum?

a) (m+2)/m = 1 b) m - 18 c) (m+2)/m = 100 d) (m+2)/m = 0.3

21. A large peak is very often found in all mass spectra at m/e = 28. What is the likely source of this peak?

a) $H_2C{=}CH_2$ b) CO c) HCN d) N_2

22. Which of the following fragmentation ions is observed from a McLafferty rearrangement?

a) $CH_3C\equiv\overset{\oplus}{O}$ b) $\overset{\oplus\,\cdot}{CH_2{=}CH_2}$ c) $\overset{\oplus}{CH_3}$ d) $CH_3\overset{\oplus}{C}HCH{=}CH_2$

23. Compounds A ($C_5H_{10}S$) and B ($C_6H_{14}O$) both show a molecular ion peak at 102 m/e. Where in the mass spectra do you look to distinguish between the two compounds?

a) m+1 b) m - 18 c) m+2 d) m - 34

24. Compound A ($C_5H_{12}O$) has a mass spectrum with no peak higher than 70 and the base peak is at 45. What is the most likely structure for compound A?

a) $CH_3(CH_2)_4OH$

b) 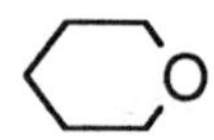

c) $CH_3(CH_2)_2OCH_2CH_3$

d) $CH_3\underset{OH}{\underset{|}{C}}HCH_2CH_2CH_3$

25. What is the species that gives a peak at m -17 for acetic acid?

a) $\overset{\oplus}{C}H_2{-}CH{=}O$ b) $CH_3C\equiv\overset{\oplus\,\cdot}{O}$ c) $CH_3{-}\overset{\oplus}{C}{=}O$ d) $CH_3{-}\underset{\cdot}{\overset{\oplus}{C}}{=}O$

Answers

1. b)
2. b)
3. c)
4. b)
5. c)
6. a)
7. b)
8. c)
9. a)
10. a)
11. d)
12. a)
13. d)
14. d)
15. c)
16. c)
17. b)
18. d)
19. b)
20. d)
21. d)
22. b)
23. c)
24. d)
25. c)

Chapter 13 Nuclear Magnetic Resonance Spectroscopy

1. What is the index of hydrogen deficiency for a compound having the molecular formula of C_3H_6O?

   a) 0 b) 1 c) 2 d) 3

2. What is the index of hydrogen deficiency for a compound having the molecular formula of $C_6H_6N_2O$?

   a) 3 b) 4 c) 5 d) 6

3. What is the index of hydrogen deficiency for a compound having a molecular formula of $C_2HO_2Cl_3$?

   a) 0 b) 1 c) 2 d) 3

4. How many sets of equivalent hydrogen atoms are there for 2-propanol?

   a) 2 b) 3 c) 4 d) 8

5. How many sets of equivalent hydrogen atoms are there for the following compound?

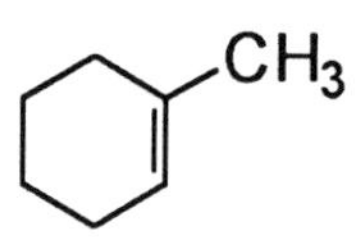

   a) 3 b) 4 c) 5 d) 6

6. How many sets of signals are there in the ^{1}H-NMR spectrum for chlorocyclohexane?

   a) 2 b) 3 c) 4 d) 6

7. How many sets of signals would you expect to see in the ^{1}H-NMR spectrum for diisopropyl ether?

   a) 1 b) 2 c) 3 d) 6

8. A radiofrequency of 75.5 MHz is required to spin-flip a carbon-13 nucleus in an applied field of 7.05 T. What is the energy in cal/mol associated with this transition?

   a) 0.00572 b) 0.00218 c) 0.0286 d) 0.00715

9. A compound gave two sets of signals in the ^{1}H-NMR spectrum with an integration ratio of 60 to 10. What is the most likely structure for the compound?

a) $CH_3CH_2CH_2\overset{\overset{O}{\|}}{C}CH_2CH_2CH_3$

b) 

c) 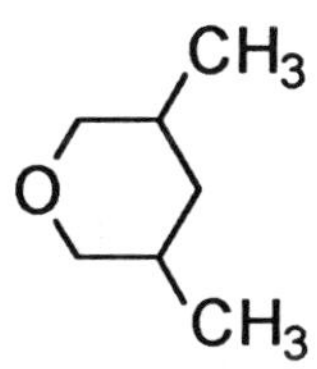

d) 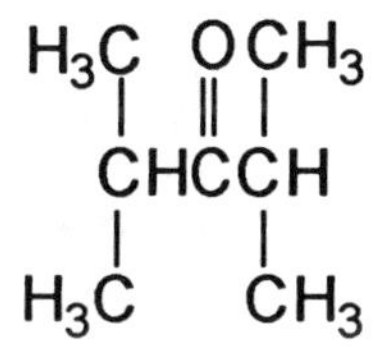

10. What are the relative areas of integration in the ^{1}H-NMR spectrum for 3-chloro-2-methylpentane?

a) 9:2:1:1 b) 3:3:3:2:1:1 c) 6:3:2:1:1 d) 6:3:2:2:1

11. Arrange the following methyl groups in the order of increasing δ values downfield from TMS in the ^{1}H-NMR spectrum (lowest first).

CH_3-NH_2	CH_3-I	CH_3-OH	CH_3-Cl
↑	↑	↑	↑
I	II	III	IV

a) I, III, IV, II b) II, I, IV, III c) IV, II, III, I d) III, IV, II, I

12. What is the order of increasing chemical shift values in the ^{1}H-NMR spectrum for the indicated hydrogen atoms (lowest first)?

$$CH_3-CH=CH-CH_2-\overset{\overset{O}{\|}}{C}OH$$

(arrows: I at CH_3, II at CH, III at CH_2, IV at OH)

a) IV, II, III, I b) I, III, II, IV c) III, I, IV, II d) II, IV, I, III

13. What is the splitting pattern for indicated hydrogen atom?

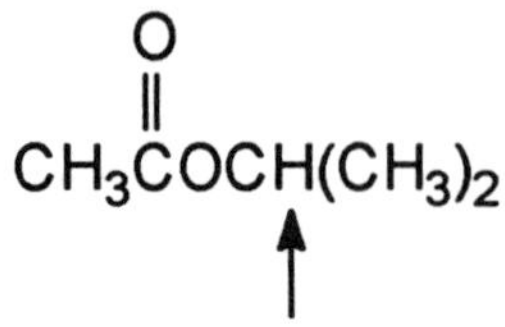

a) doublet b) triplet c) quartet d) septet

14. What are the splitting patterns for the hydrogen atoms in 2,2-dibromopropane?

a) septet b) quartet c) doublet d) singlet

15. The splitting pattern for compound A, C_3H_5OCl, gave a quartet and a triplet. Which of the following structures is most likely compound A?

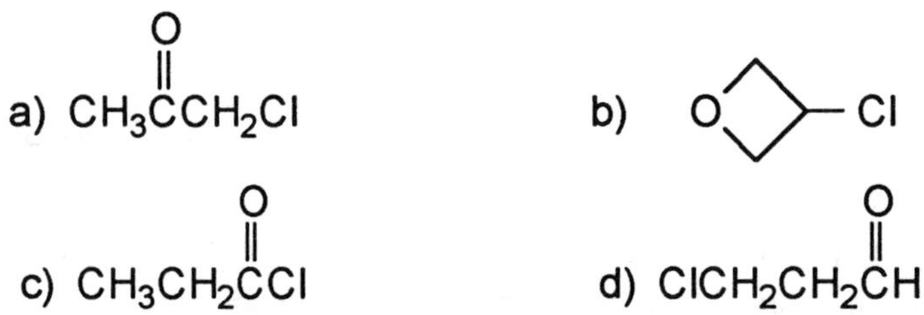

16. The splitting pattern for compound A, C_3H_5OCl, gave a quintet and a doublet. Which of the following structures is most likely compound A?

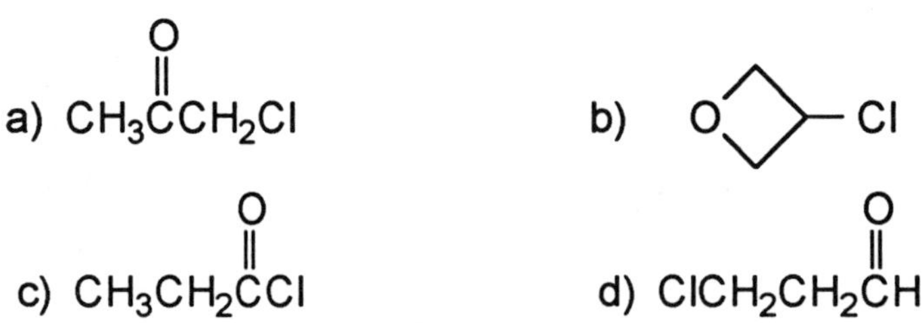

17. Which of the following compounds show spin spin splitting in the ^{1}H-NMR spectrum?

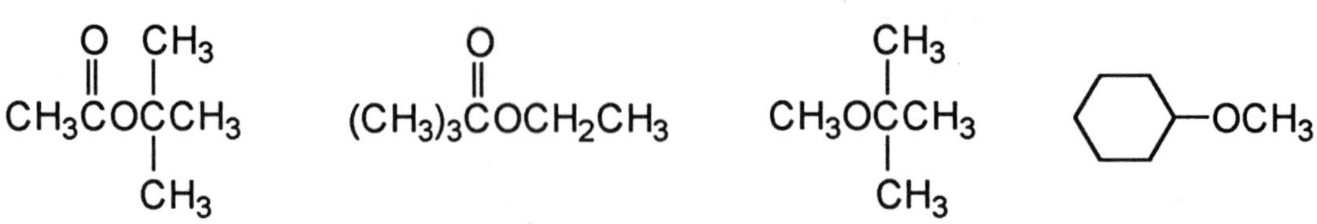

I II III IV

a) I, II b) III, IV c) I, III d) II, IV

18. Which of the following compounds give a singlet with 3 hydrogens at about 2.0 ppm in the ^{1}H-NMR spectrum?

a) 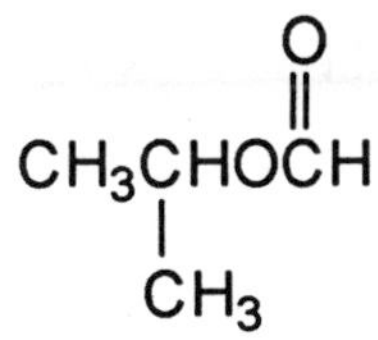

b) 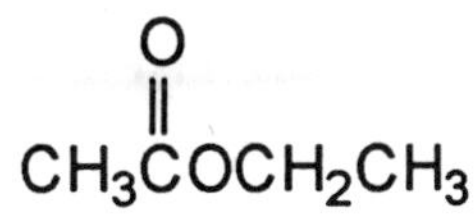

c) 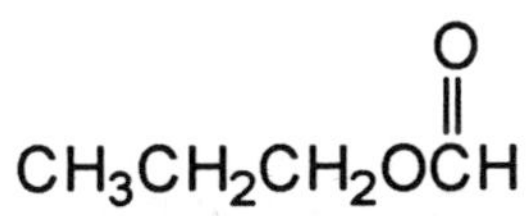

d) 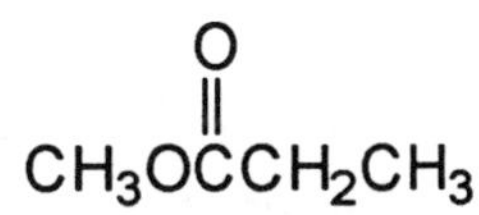

19. Which of the following compounds give a singlet with 3 hydrogens at about 4.0 ppm in the ^{1}H-NMR spectrum?

a) 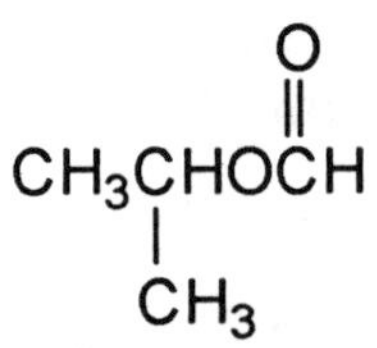

b) 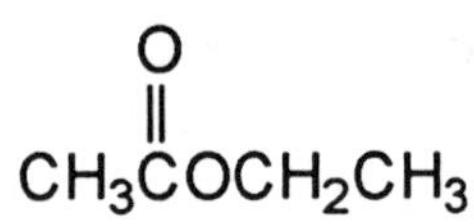

c) 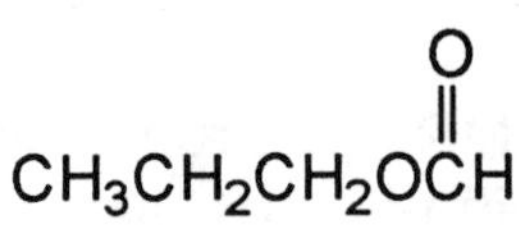

d) 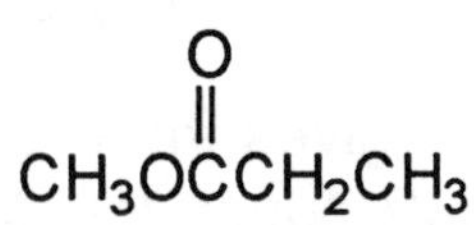

20. Which of the following compounds gives four sets of signals in the ^{1}H-NMR spectrum?

a) 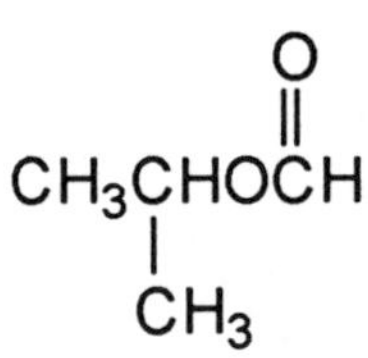

b) 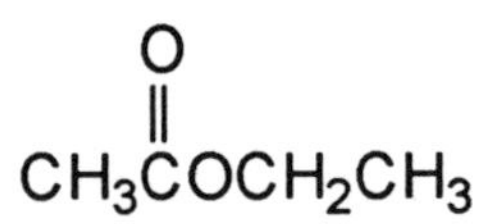

c) 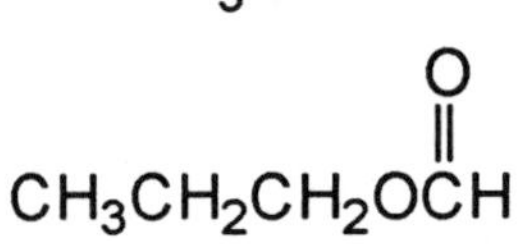

d) 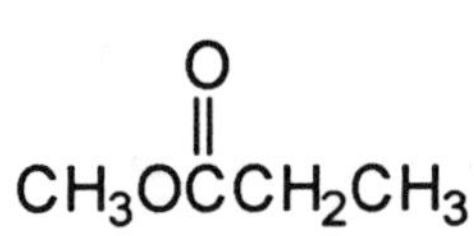

21. Which of the following can be used to distinguish between pentane and 2-methylbutane in the ^{13}C NMR spectrum?

a) The number of signals
b) The proton coupled C NMR spectra
c) The chemical shifts
d) All of these

22. What is the order of increasing δ values in the ^{13}C NMR spectra for the following carbon atoms?

CH_4	CH_3F	CH_3Br	CH_3OH	CH_3Cl
↑	↑	↑	↑	↑
I	II	III	IV	V

a) III, V, II, IV, I b) I, IV, II, V, III c) V, IV, II, III, I d) I, III, V, IV, II

23. Compound A gives a single signal in the ^{13}C NMR spectrum and a single signal in the ^{1}H-NMR spectrum. Which of the following is most likely compound A?

a) dimethyl ether b) diethyl ether c) neopentane d) methyl acetate

24. Compound B gives two signals in the ^{13}C NMR spectrum and a single signal in the ^{1}H-NMR spectrum. Which of the following is most likely compound B?

a) dimethyl ether b) diethyl ether c) neopentane d) methyl acetate

25. Compound C gives three signals in the ^{13}C NMR spectrum and two signals in the ^{1}H-NMR spectrum. Which of the following is most likely compound C?

a) dimethyl ether b) diethyl ether c) neopentane d) methyl acetate

Answers

1. b)
2. c)
3. b)
4. a)
5. d)
6. c)
7. b)
8. d)
9. d)
10. c)
11. b)
12. b)
13. d)
14. d)
15. c)
16. b)
17. d)
18. b)
19. d)
20. c)
21. d)
22. d)
23. a)
24. c)
25. d)

Chapter 14 Infrared and UV-Visible Spectroscopy

1. Which spectroscopic method is most useful for quantitative determinations?

   a) NMR b) Mass Spec c) IR d) UV-Vis

2. What is the approximate energy range for absorptions in the infrared region of the electromagnetic spectrum?

   a) 50 to 100 kcal/mol b) 10 to 50 kcal/mol
   c) 1 to 10 kcal/mol d) 0.01 to 0.1 kcal/mol

3. What is the wave number for an infrared band at 5 microns?

   a) 500 cm^{-1} b) 2000 cm^{-1} c) 5000 cm^{-1} d) 10,000 cm^{-1}

4. What is the wavelength (μm) of an infrared absorption band at a wavenumber of 500 cm^{-1}?

   a) 0.2 b) 2.0 c) 20 d) 25

5. Place the following species in the order of increasing bond stretching frequencies (lowest first).

C=N	C=O	C=C	C=S
I	II	III	IV

   a) II, I, IV, III b) I, II, IV, III c) III, I, II, IV d) IV, III, I, II

6. The broadening of the stretching vibration peak for alcohols is mainly due to which of the following?

   a) water contamination b) a strong dipole moment
   c) hydrogen bonding d) coupling peaks

7. A compound has the molecular formula, $C_5H_{10}O$, and strong infrared absorptions at 1100 and 3350 cm^{-1}. Which of the following is the most likely structure for the compound?

   a) 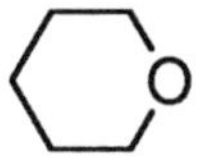

   b) $CH_3CH_2\overset{\overset{O}{\|}}{C}CH_2CH_3$

   c) 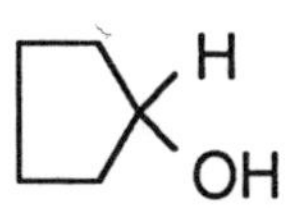

   d) $CH_3CH{=}CHCH_2OCH_3$

8. Which of the following compounds has a sharp IR absorption band at 1710 cm^{-1} and a broad band at 3300 cm^{-1}?

a) CH_3COOH b) CH_3CH_2OH c) $CH_3C(=O)CH_3$ d) CH_3OCH_3

9. Which of the following compounds has an IR absorption band at 3300 cm^{-1} but no band at 1710 cm^{-1}?

a) CH_3COOH b) CH_3CH_2OH c) $CH_3C(=O)CH_3$ d) CH_3OCH_3

10. What is the most likely structure for compound X that has the molecular formula, C_5H_8O, and an IR spectrum with a peak at 1710 cm^{-1}, but no peak at 1650 cm^{-1}?

a) 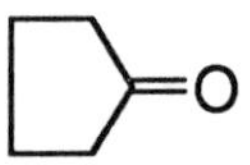 b) 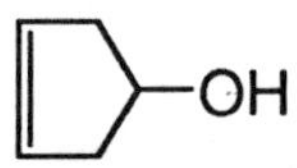 c) 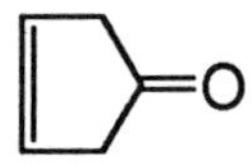 d) $CH_3CH{=}CHCH_2CHO$

11. Which of the following compounds has the characteristic IR peaks at 3300 cm^{-1} and at 2120 cm^{-1}?

a) $CH_3CH{=}CHCH(CH_3)CH_3$ b) $CH{\equiv}CCH_2CH_3$

c) $CH_3CH_2CH_2OCH_2CH_3$ d) $CH_3C(CH_3)_2CH_2CH_3$

12. Which of the following compounds has the characteristic IR peaks at 3100 cm^{-1} and at 1650 cm^{-1}?

a) $CH_3CH{=}CHCH(CH_3)CH_3$ b) $CH{\equiv}CCH_2CH_3$

c) $CH_3CH_2CH_2OCH_2CH_3$ d) $CH_3C(CH_3)_2CH_2CH_3$

13. Which of the following compounds has the characteristic IR peak at 3300 cm^{-1}?

a) 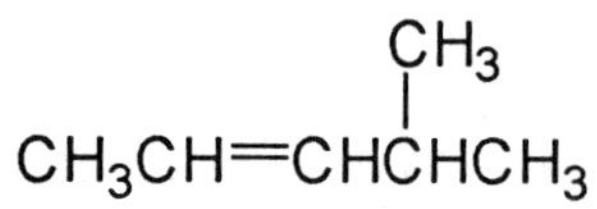

b) $CH{\equiv}CCH_2CH_3$

c) $CH_3CH_2CH_2OCH_2CH_3$

d) 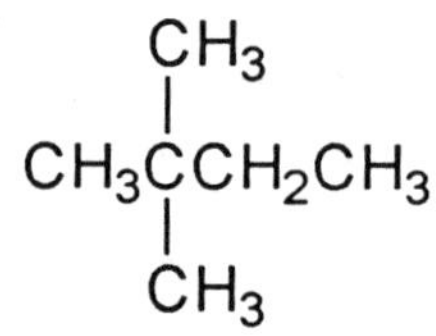

14. In which of the following will the characteristic stretching frequency in the IR spectra be particularly weak?

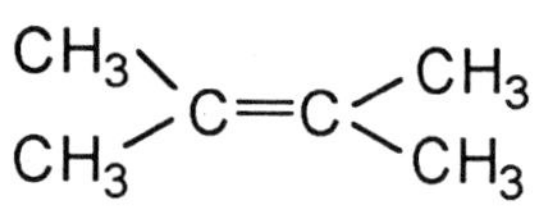

I

$CH_3C{\equiv}CCH_3$

II

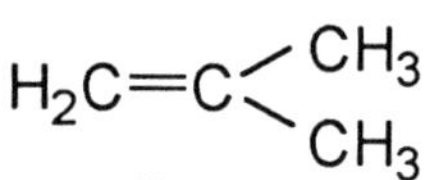

III

$CH_3CH_2C{\equiv}CH$

IV

a) I, II   b) III, IV   c) I, III   d) II, IV

15. UV-visible spectroscopy is used mostly to detect which of the following transitions?

a) electronic   b) molecular vibrational
c) dipole changes   d) symmetrical changes

16. What is the order of increasing energy for the following transitions (lowest first)?

I. π -------> π^* II. n -------> π^* III. n -------> σ^*

a) I, III, II   b) II, I, III   c) III, II, I   d) I, II, III

17. The extinction coefficient of a compound having the molecular weight of 107 is 13,100. How many milligrams of this material must be dissolved in 1 mL to give an absorbance of 0.16 in a light path of 1.00 cm?

a) 1.3   b) 131   c) 0.13   d) 22

18. The absorbance of a 1 mL sample of double-stranded DNA, which has the extinction coefficient of 6670 at 260 nm, is 0.5128 in a 1 cm cuvette. If the molecular weight is taken as 650 for one repeating unit in the polymer, how many mg are in the sample?

a) 5 b) 50 c) 75 d) 100

19. Which of the following compounds have an absorption in the UV greater than 200 nm?

I. 1,2-pentadiene II. 1,3-pentadiene III. 1,4-pentadiene
IV. 2,3-pentadiene V. 2,4-hexadiene

a) I, II b) II, V c) I, V d) III, IV

20. Which of the following compounds has an absorption maximum at 226 nm?

a) 1,3-cyclohexadiene
b) 1,4-cyclohexadiene
c) 1,4-hexadiene
d) 1,5-hexadiene

21. What is the approximate energy range for absorptions in the UV?

a) 0.01-0.1 kcal/mol
b) 1-10 kcal/mol
c) 10-50 kcal/mol
d) 50-100 kcal/mol

22. A compound has the molecular weight of 82 and a UV maximum at about 220. The proton NMR has 3 sets of signals and the IR has a band at 3030 and another at 1680. What is the most likely structure for the compound?

a) $CH_3CH_2C{\equiv}CCH_2CH_3$

b) 

c) 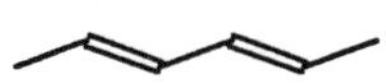

d) 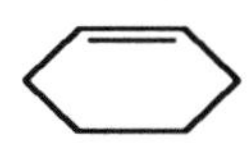

23. Compound X has the molecular formula C_4H_8O. The NMR has a doublet of 3 hydrogen atoms at 1.5 ppm and 3 hydrogens in the chemical shift range of 5-6 ppm (in addition to other signals). The IR has bands at 3400 and 1650.

a) 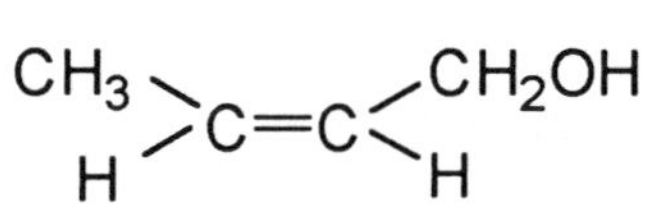

b) 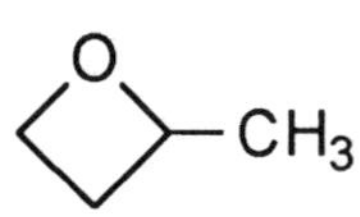

c) $CH_3CH(OH)CH{=}CH_2$

d) $CH_3OCH_2CH{=}CH_2$

24. Compound wow absorbs in the visible region at the wave length of 575 - 580. What color is it?

a) red b) yellow c) blue d) green

25. One of the best known color indicators is phenolphthalein, which is colorless at low pH and red at high pH. What is the best explanation for the change?

a) hydrogen bonding at low pH
b) reaction to form high energy species
c) molecular rearrangements
d) reaction to form multiple conjugated aromatic rings

Answers

1. d)
2. c)
3. b)
4. c)
5. d)
6. c)
7. c)
8. a)
9. b)
10. a)
11. b)
12. a)
13. b)
14. a)
15. a)
16. b)
17. a)
18. a)
19. b)
20. a)
21. d)
22. c)
23. c)
24. c)
25. d)

Chapter 15 Aromatics I

1. Which of the following species is aromatic?

a) b) c) d)

2. Which of the following compounds is antiaromatic?

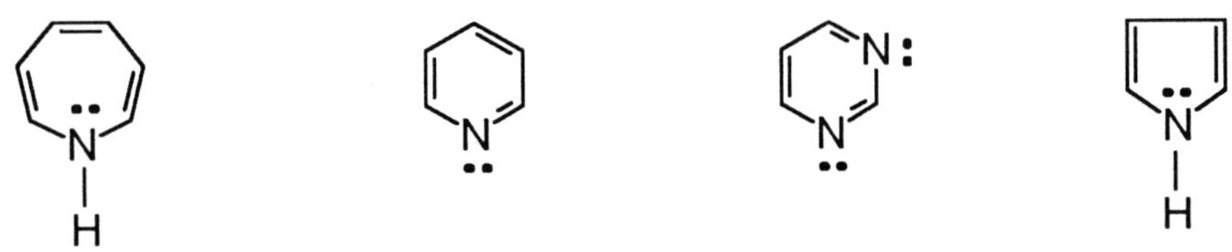

a) b) c) d)

3. Loss of a proton leads to an aromatic species for which of the following compounds?

I II III IV

a) I, II b) III, IV c) I, III d) II, IV

4. Loss of a hydride leads to an aromatic species for which of the following compounds?

I II III IV

a) I, II b) III, IV c) I, III d) II, IV

5. Which of the following compounds is the strongest acid?

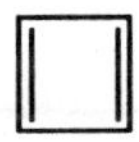

a) b) c) d)

6. Which of the following structures have the correct names?

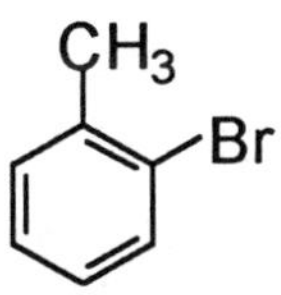

I. 2-bromotoluene

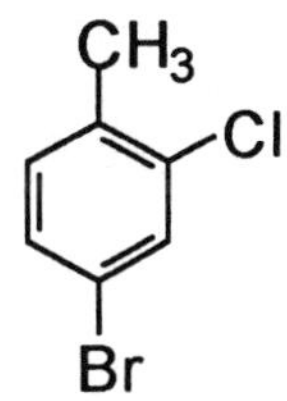

II. para bromo, ortho chloro toluene

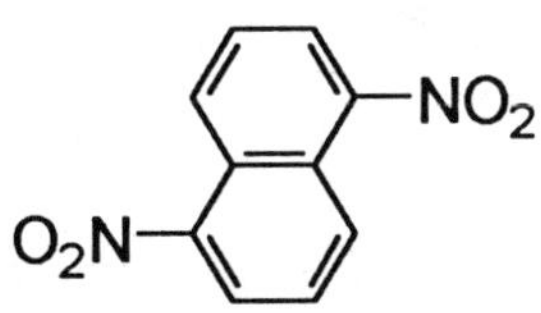

III. 1,6-dinitronaphthalene

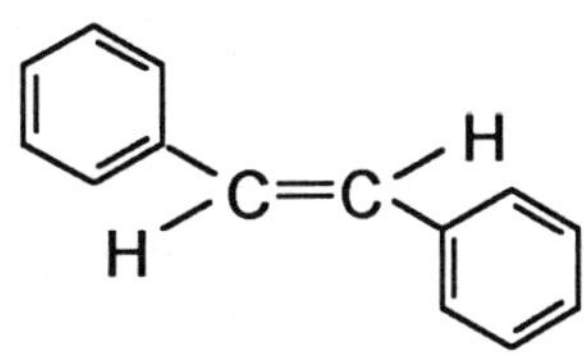

IV. trans stilbene

a) I, II b) III, IV c) I, IV d) II, III

7. What is the correct name for the following structure?

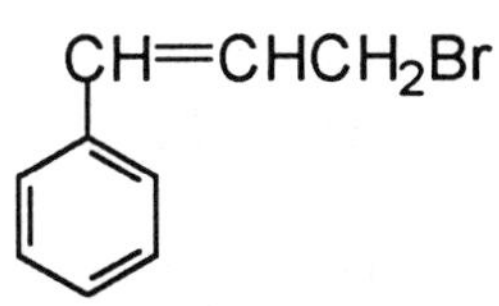

a) α-styryl bromide
b) β-styryl bromide
c) cinnamyl bromide
d) 1-bromo-3-phenylpropene

8. How many carbon resonances are there in the ^{13}C NMR spectrum for the following compound?

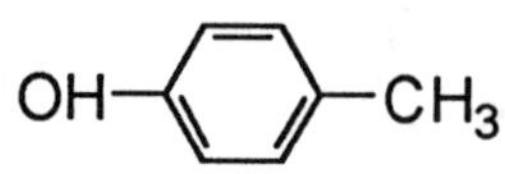

a) 3 b) 4 c) 5 d) 6

9. Compound X has a mass spectrum containing, among other peaks, the molecular ion peak at 134 (100%) and a peak at 135 (11%). This compound does not form a carboxylic acid on vigorous oxidation. What is the most likely structure for compound X?

a) 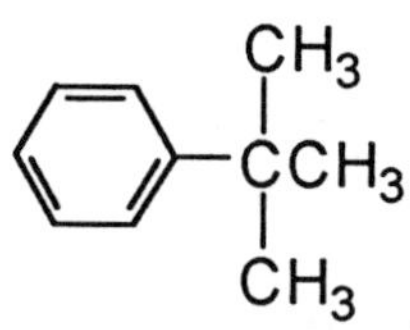

b) 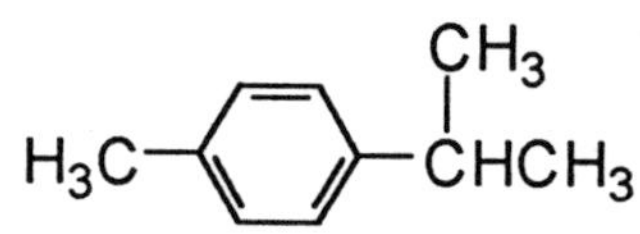

c) 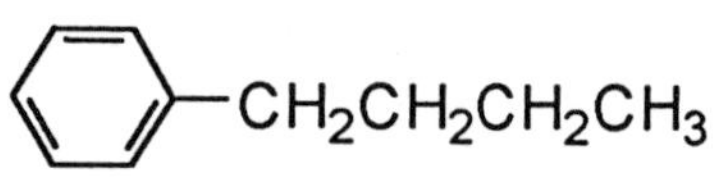

d) 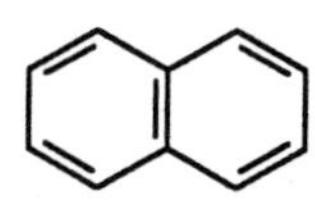

10. What is the order of increasing acidity for the following compounds (least first)?

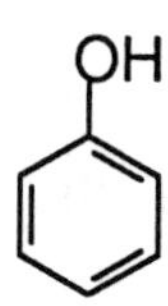

I

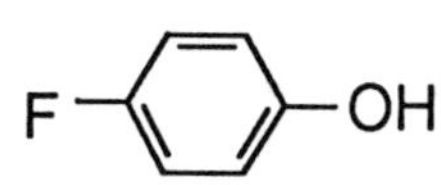

II

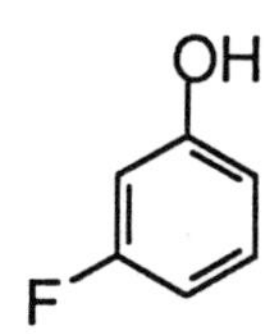

III

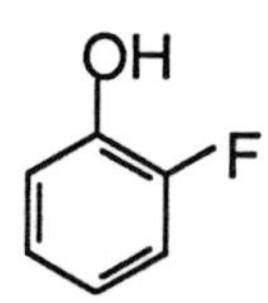

IV

a) I, II, III, IV b) III, II, IV, I c) I, IV, III, II d) II, III, IV, I

11. What is the order of increasing acidity for the following compounds (least first)?

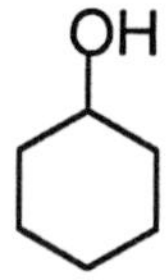

I

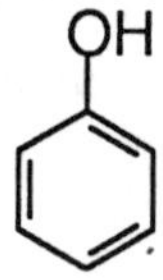

II

H_2CO_3

III

CH_3CO_2H

IV

a) I, II, III, IV b) III, II, IV, I c) I, IV, III, II d) II, III, IV, I

12. Compound A is heated to give the following compound. What is a reasonable structure for compound A?

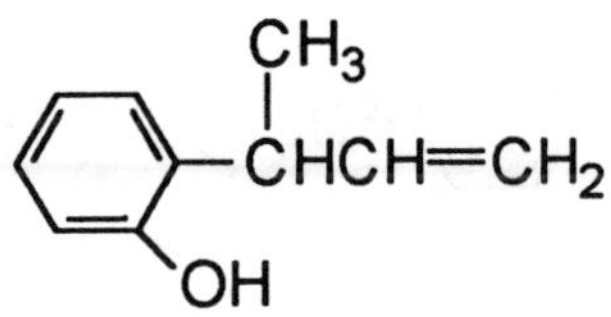

a) 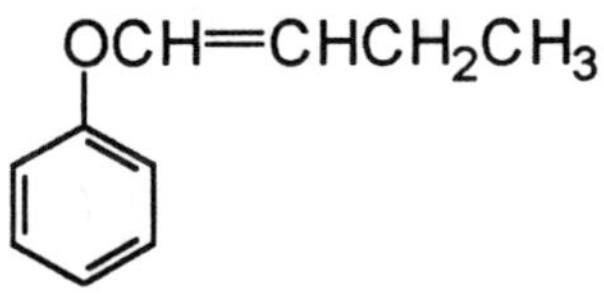

b) 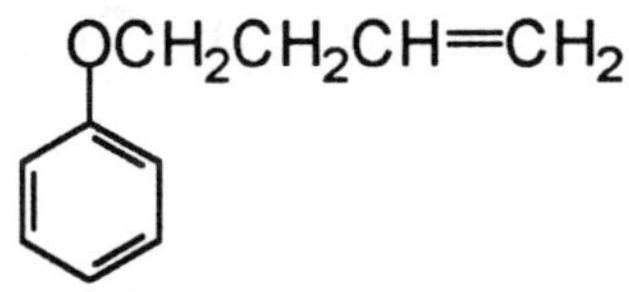

c) 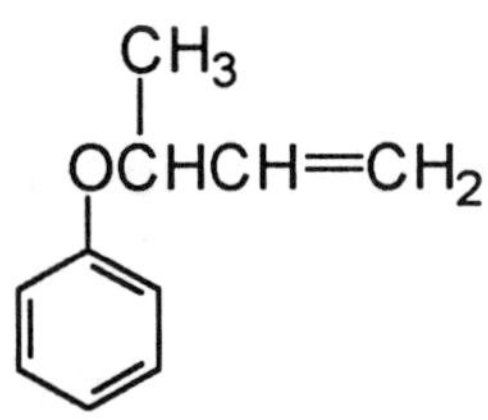

d) 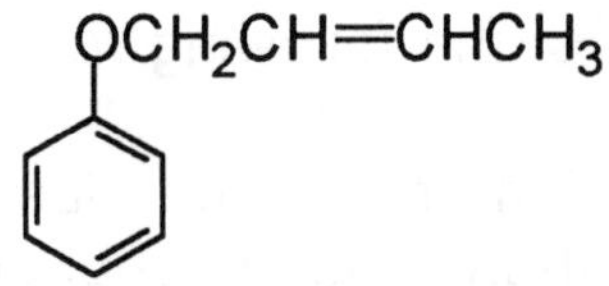

13. What is the most likely product from the reaction of ethyl benzene with bromine, followed by reaction with potassium methoxide in methanol?

a) 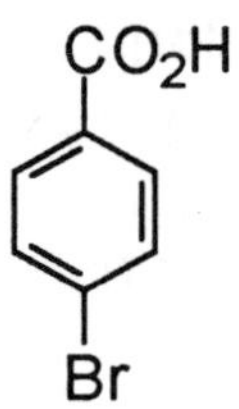

b) 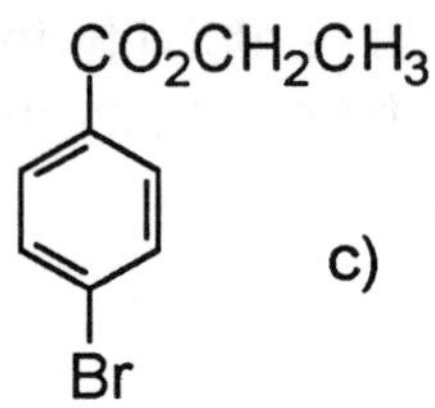

c) 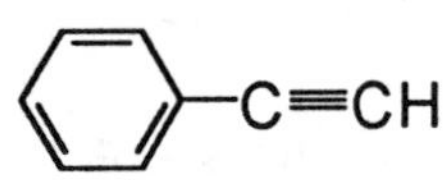

d) 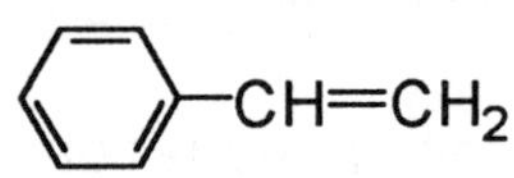

14. What is the most likely product from the following series of reactions?

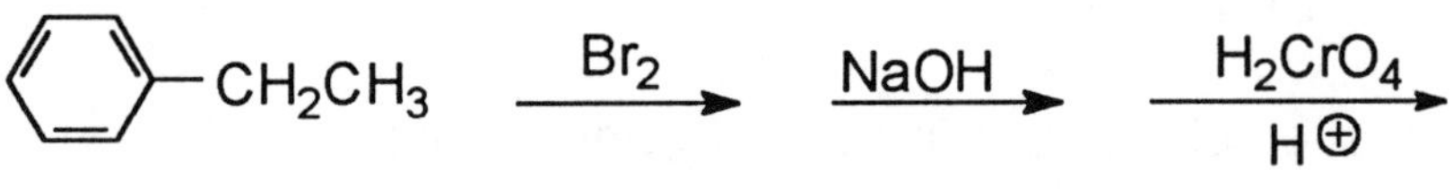

a) 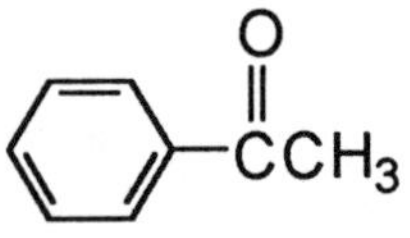

b) 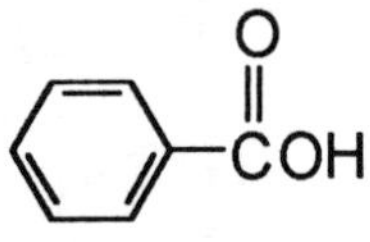

c) 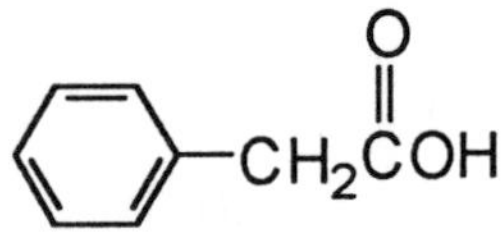

d) 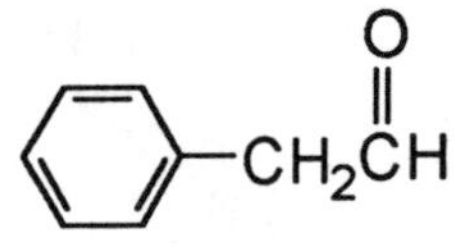

15. What is the product from the following reaction?

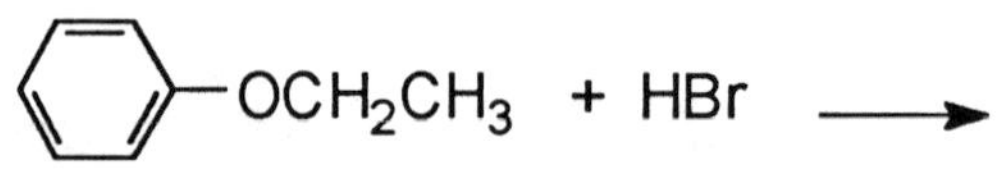

a) 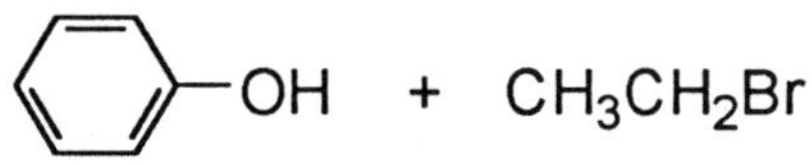

b) 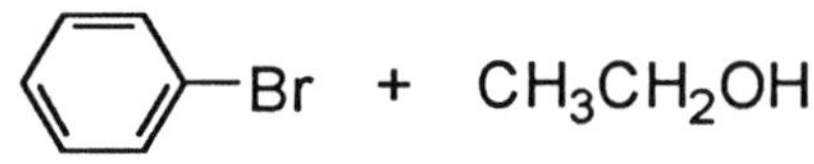

c) 

d) 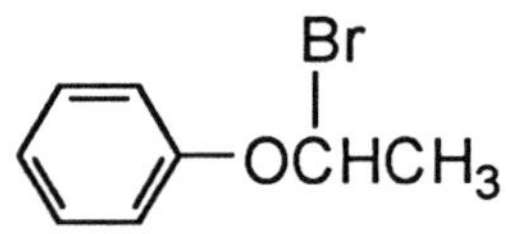

16. Which of the following reactions is the best procedure for preparing anisole?

a) sodium methoxide plus bromobenzene
b) methanol plus phenol plus acid
c) phenol plus acid followed by methyl bromide plus acid
d) phenol plus base followed by methyl bromide

17. How can phenol be distinguished from cyclohexanol?

a) solubility in water
b) solubility in hydrochloric acid
c) solubility in sodium bicarbonate
d) solubility in sodium hydroxide

18. How many resonance structures are possible for the benzyl cation?

a) 3 b) 4 c) 5 d) 6

19. How many Kekule´ structures are possible for naphthalene?

a) 3 b) 4 c) 5 d) 6

20. How many p-orbital electrons are in the following molecule?

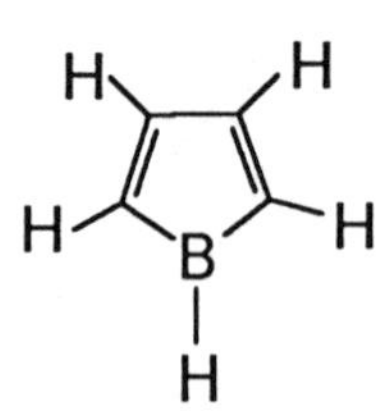

a) 4 b) 6 c) 8 d) 10

21. How many p-orbital electrons are in the following molecule?

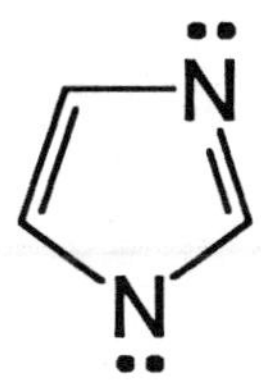

a) 4 b) 6 c) 8 d) 10

22. What is the best way to distinguish between benzyl alcohol and ortho cresol?

a) mass spectrometry
b) IR spectroscopy
c) UV-Visible spectroscopy
d) proton NMR spectroscopy

23. How many mono bromoanthracenes are possible?

a) 1 b) 2 c) 3 d) 4

24. How many nodes are there in the highest occupied molecular orbital of benzene?

a) 0 b) 1 c) 2 d) 3

25. A compound has the molecular formula, C_8H_9Br, and a proton NMR spectrum with a multiplet at 7.2 ppm (2H), and another multiplet at 7.6 ppm (2H). What is the most likely structure for the compound?

a) 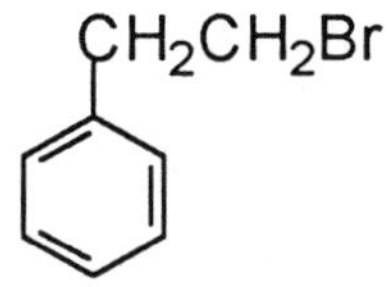

b) 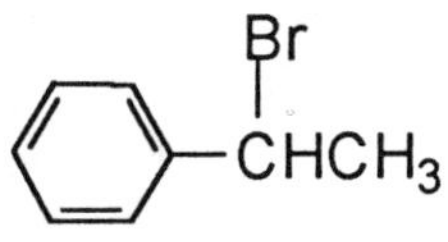

c) 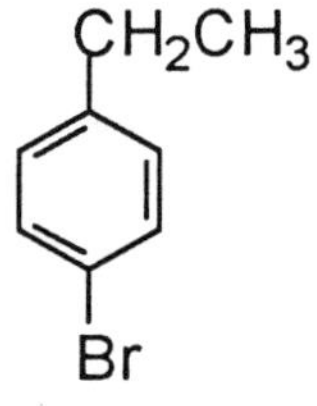

d) 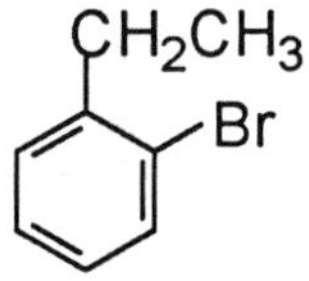

Answers

1. c)
2. a)
3. d)
4. c)
5. c)
6. c)
7. c)
8. c)
9. a)
10. a)
11. a)
12. a)
13. d)
14. a)
15. a)
16. d)
17. d)
18. b)
19. a)
20. a)
21. b)
22. d)
23. c)
24. b)
25. c)

Chapter 16 Aromatics II

1. Which of the following contributing resonance structures is the most important for the bromination of anisole?

a) 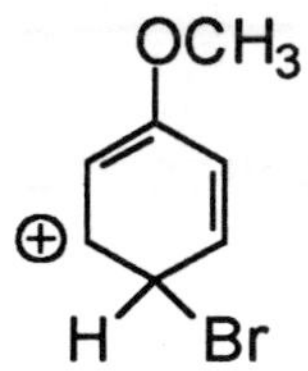 b) 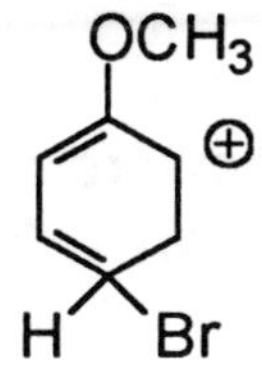 c) 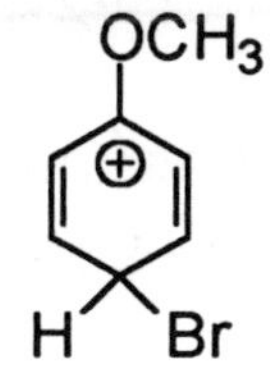 d) 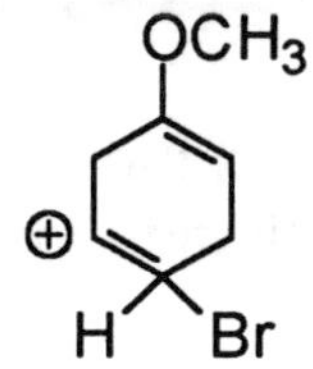

2. Which of the following groups is not a meta-director?

a) $-NO_2$ b) 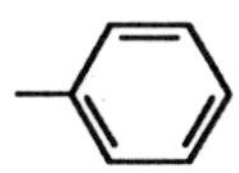 c) $-CN$ d) $-\overset{\oplus}{N}(CH_3)_3$

3. Which of the following groups are meta directors for electrophilic aromatic substitution reactions?

$-NO_2$	$-COCH_3$	$-OCH_3$	$-COOH$	$-NHCOCH_3$	$-NH_2$
I	II	III	IV	V	VI

a) II, III, IV b) III, V, VI c) I, II, IV d) I, IV, VI

4. Which of the following groups is an ortho-para director?

a) $-OCOCH_3$ b) $-COOCH_3$ c) $-CHO$ d) $-COOH$

5. Which of the following groups activate the benzene ring for electrophilic aromatic substitution?

$-Br$	$-OH$	$-OCH_3$	$-COOH$	$-\overset{\oplus}{N}H_3$	$-CH_3$
I	II	III	IV	V	VI

a) II, III, VI b) III, V, VI c) I, II, IV d) I, IV, VI

6. What is the electrophile in the nitration of benzene?

a) HNO_3 b) $\overset{\oplus}{N}O_2$ c) $\overset{\ominus}{N}O_3$ d) $\overset{\oplus}{N}O$

7. What is the order of increasing reactivity toward electrophilic aromatic substitution for the following compounds (lowest first)?

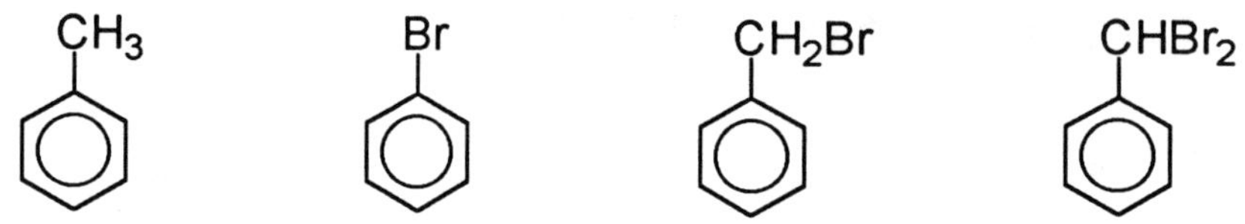

I II III IV

a) II, I, IV, III b) I, III, IV, II c) III, IV, I, II d) II, IV, III, I

8. What is the order of increasing reactivity toward electrophilic aromatic substitution for the following compounds (lowest first)?

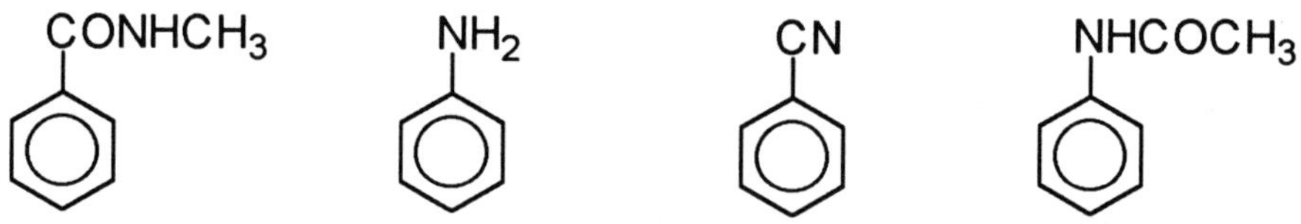

I II III IV

a) II, I, IV, III b) I, III, IV, II c) III, IV, I, II d) IV, III, II, I

9. Which compound reacts most readily in the nitration reaction?

a) 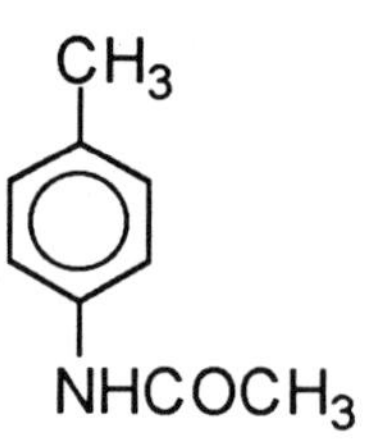

b) 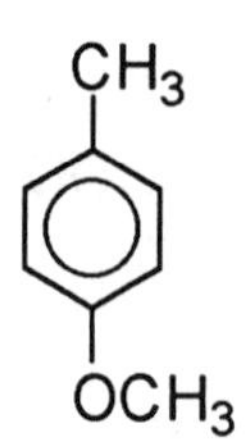

c) 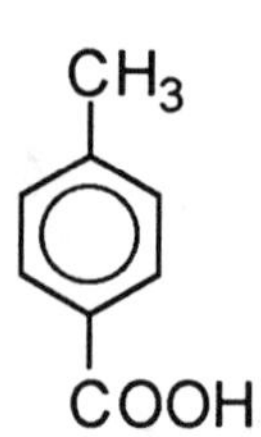

d) 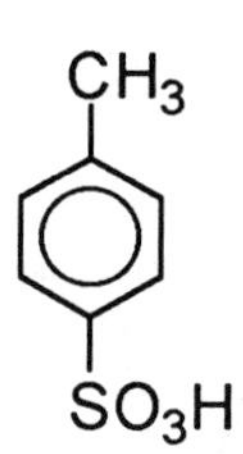

10. Which of the following is a correct resonance structure for the intermediate formed by nitration of benzene?

a) 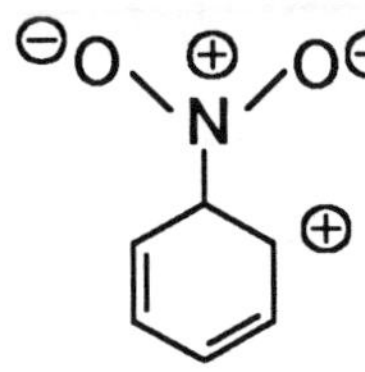

b) 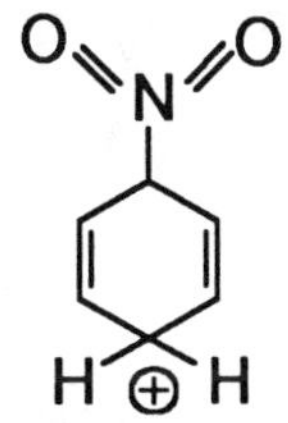

c) 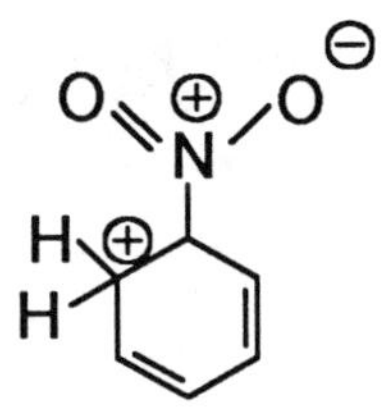

d) 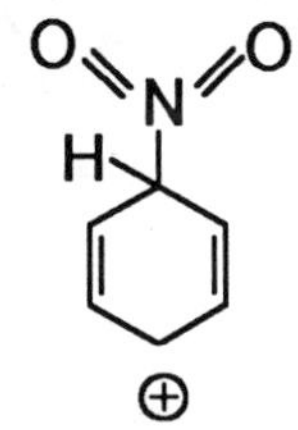

11. What is the major product from the following reaction?

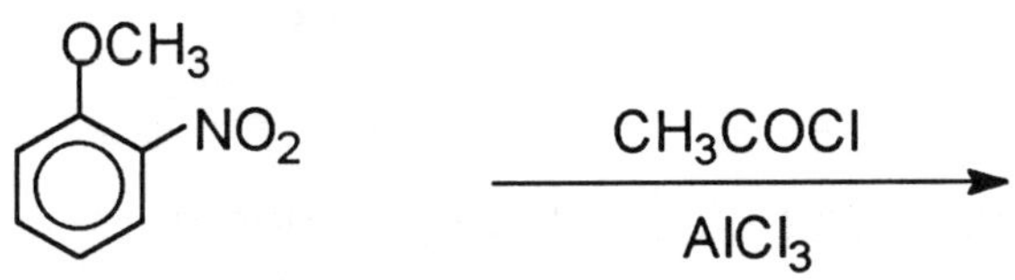

a) 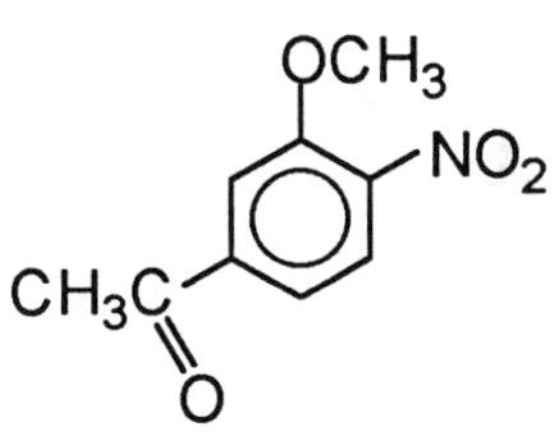

b) 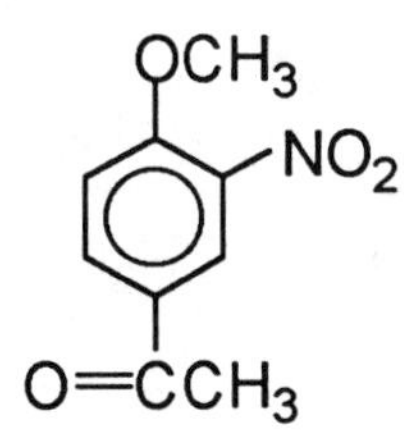

c) 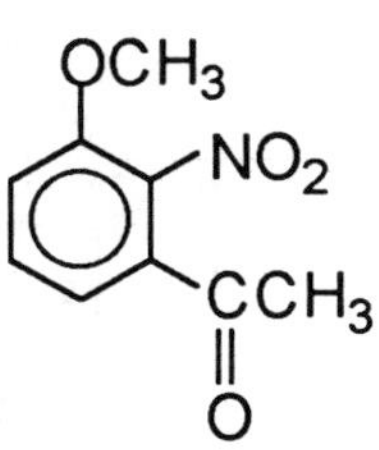

d) 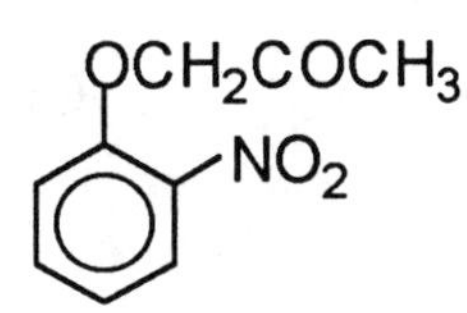

12. What is a major product from the following reaction?

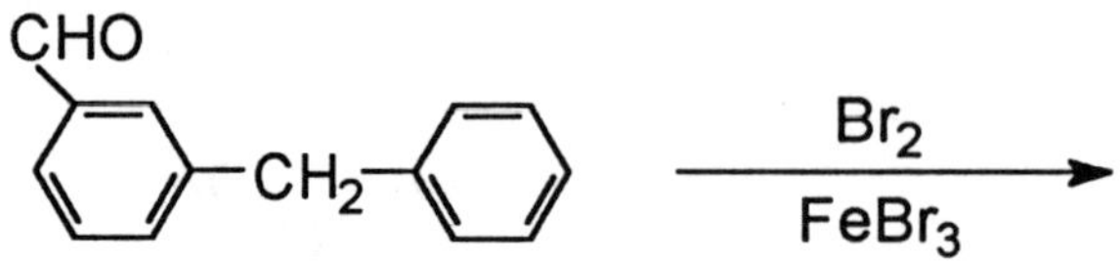

a) 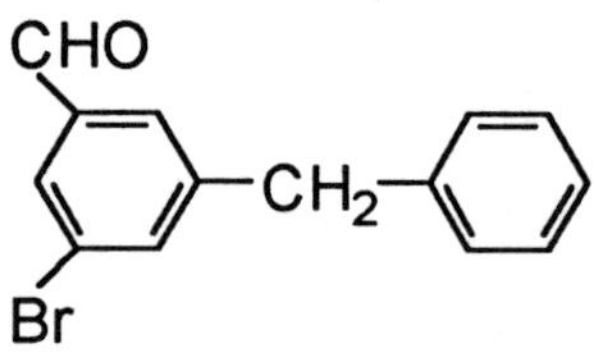

b) 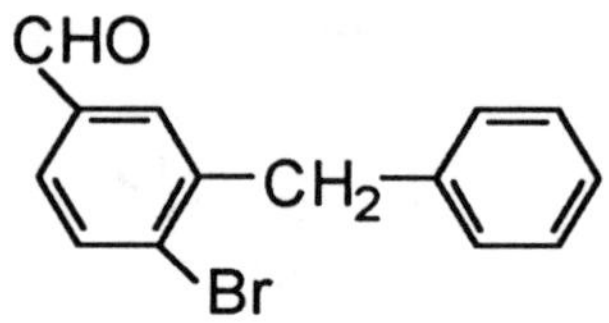

c) 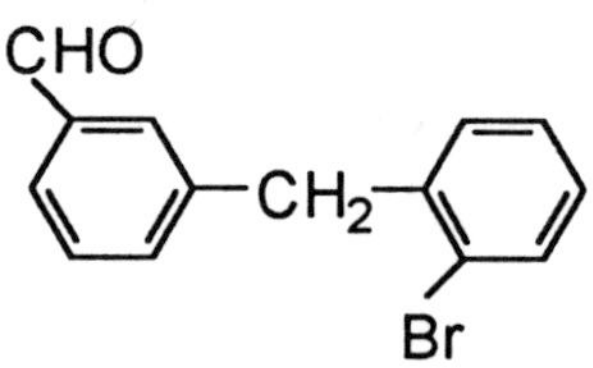

d) 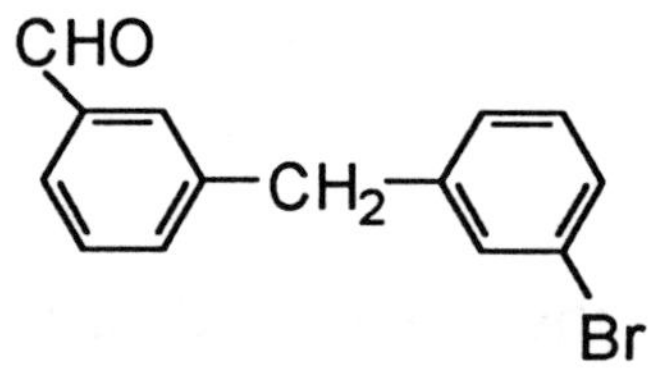

13. What is the major product from the following series of reactions?

a) 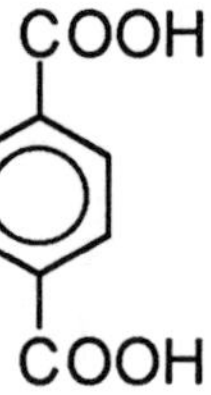

b) 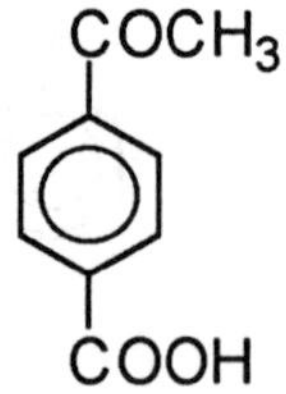

c) 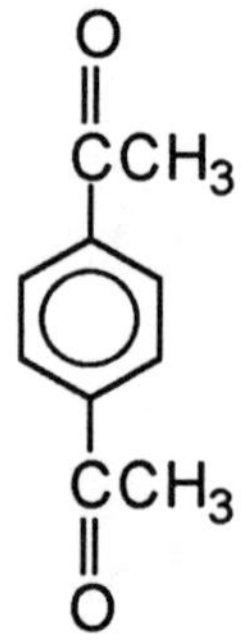

d) 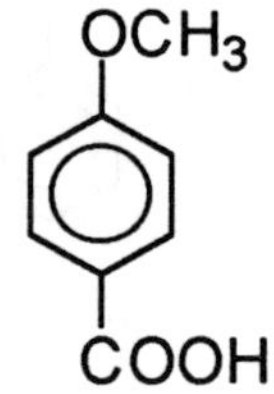

14. What is the best reaction sequence for preparing the following compound from benzene?

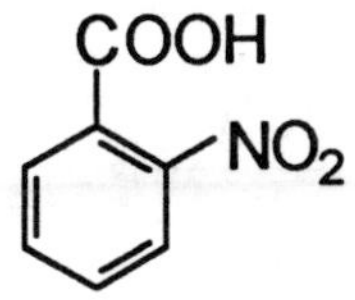

a) $\xrightarrow[CH_3Cl]{AlCl_3}$ $\xrightarrow[^{\ominus}OH\text{ , heat}]{KMnO_4}$ $\xrightarrow[H_2SO_4]{HNO_3}$

b) $\xrightarrow[H_2SO_4]{HNO_3}$ $\xrightarrow[CH_3Cl]{AlCl_3}$ $\xrightarrow[^{\ominus}OH\text{ , heat}]{KMnO_4}$

c) $\xrightarrow[CH_3OCl]{AlCl_3}$ $\xrightarrow[H_2SO_4]{HNO_3}$

d) $\xrightarrow[CH_3Cl]{AlCl_3}$ $\xrightarrow[H_2SO_4]{HNO_3}$ $\xrightarrow[^{\ominus}OH\text{ , heat}]{KMnO_4}$

15. What is the major product from the reaction of benzene with 1-chlorobutane and aluminum chloride?

a) 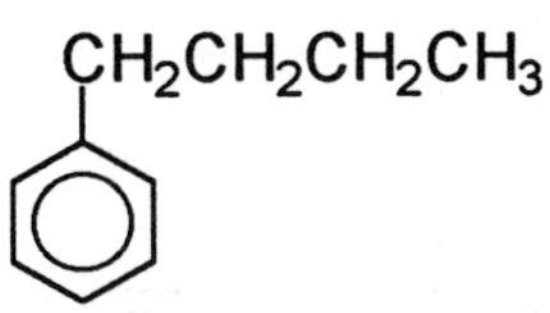

b) 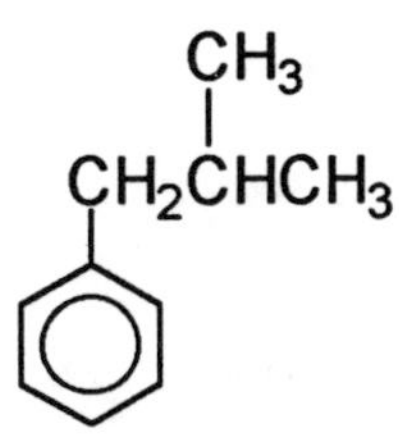

c) 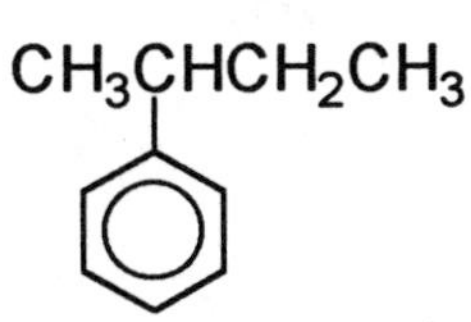

d) 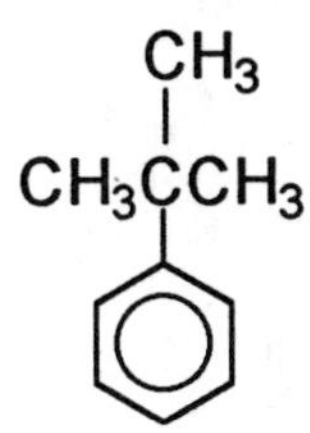

16. What is the major product from the following reaction?

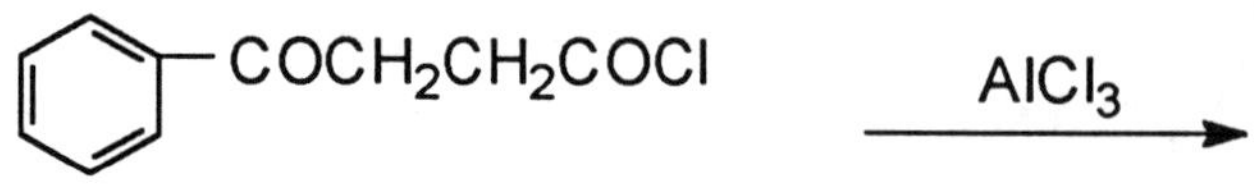

a) 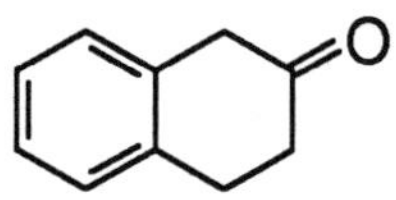 b) 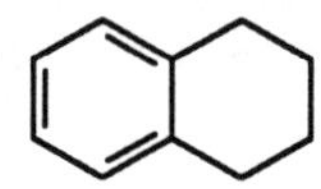

c) 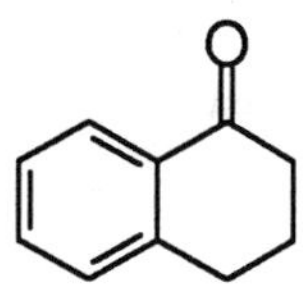 d) 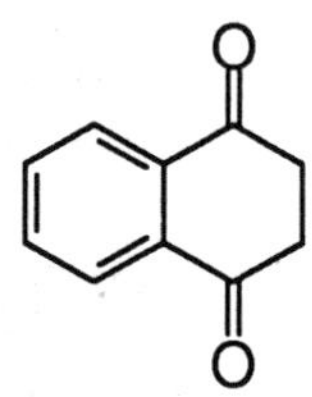

17. What is the intermediate in the final reaction step for the following reaction?

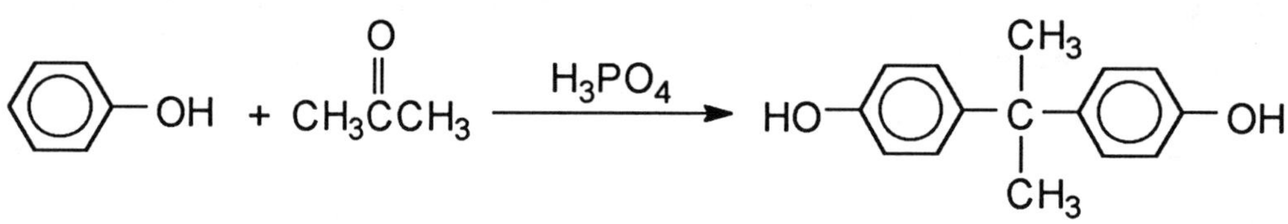

a) 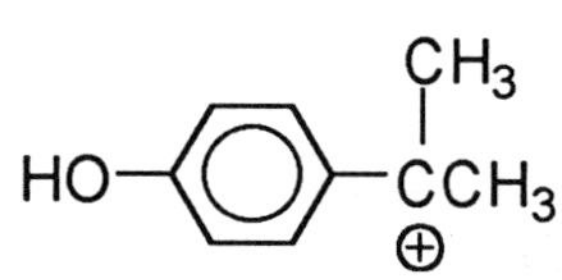 b) 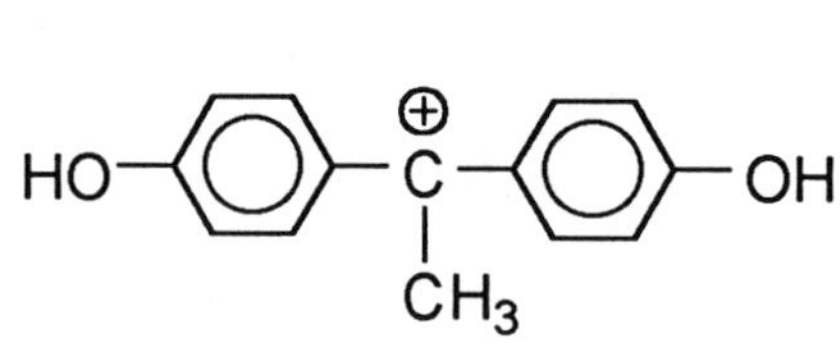

c) 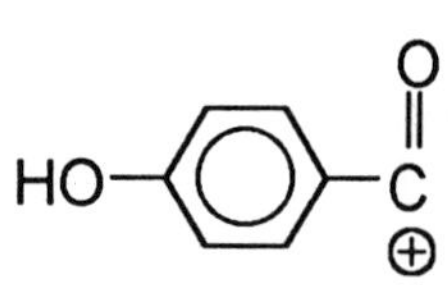 d) 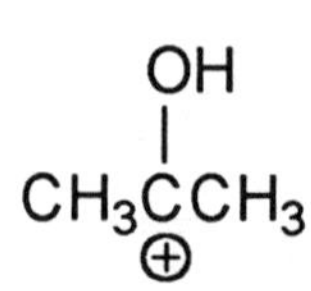

18. Which of the following groups activate a nucleophilic aromatic substitution reaction?

$-Cl$	$-OCH_3$	$-CF_3$	$-NH_2$	$-CH_3$	$-CN$
I	II	III	IV	V	VI

a) I, III, VI b) II, V, VI c) I, II, III d) IV, V, VI

19. What is the best resonance form for the sigma complex in a nucleophilic aromatic substitution reaction?

a) 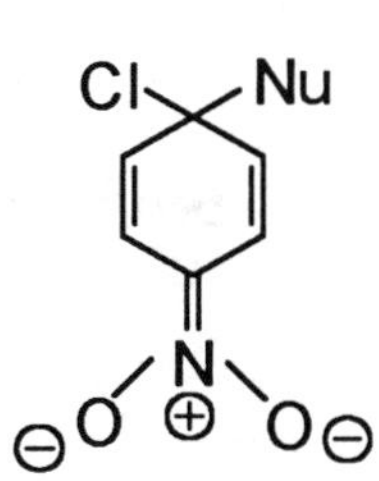

b) 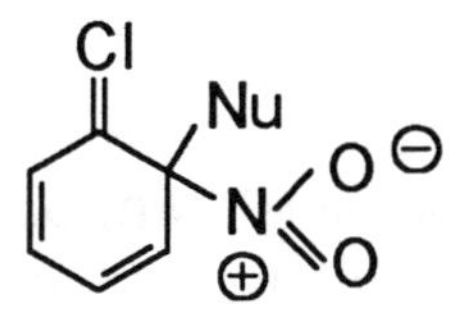

c) 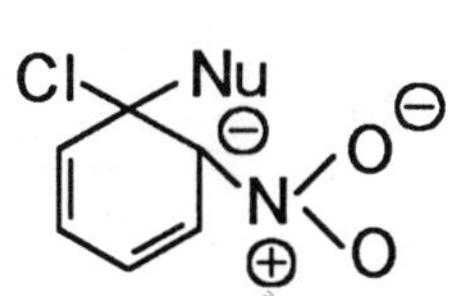

d) 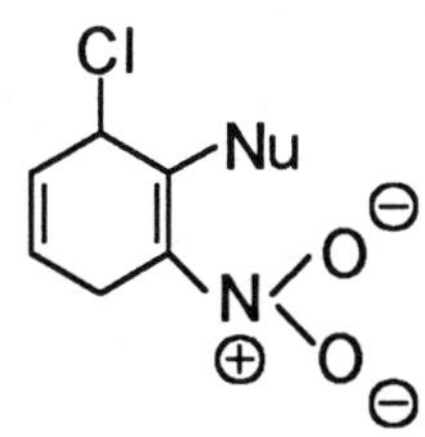

20. Which of the following compounds reacts fastest with sodium hydroxide in water?

a) 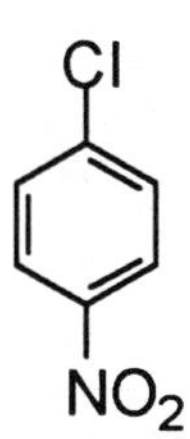

b) 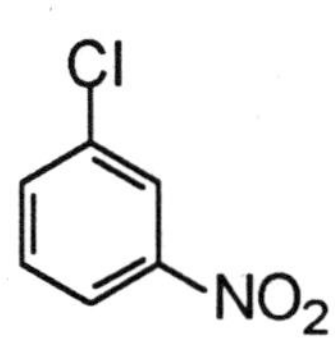

c) 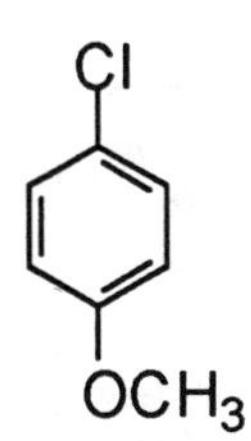

d) 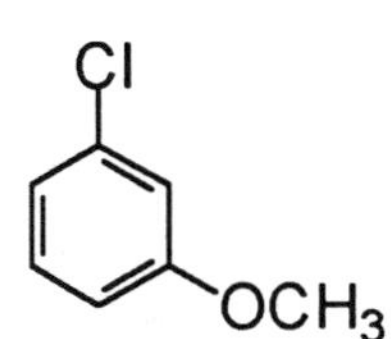

21. What is the principle product from the following reaction?

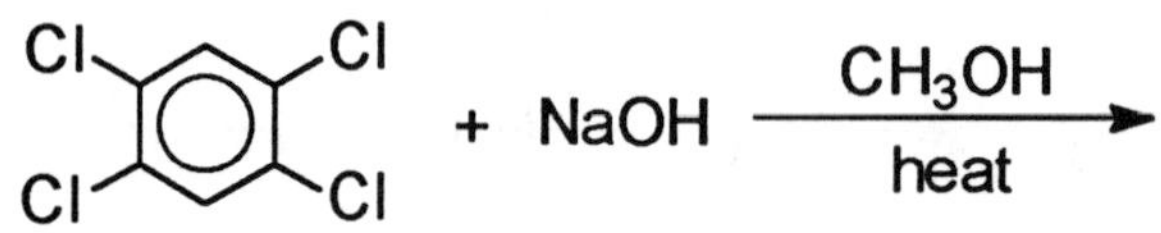

a) 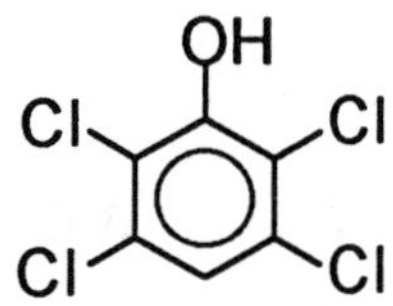

b) 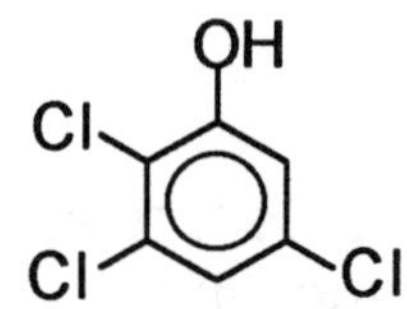

c) 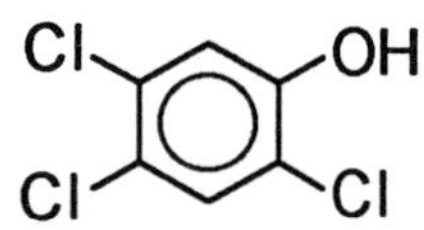

d) 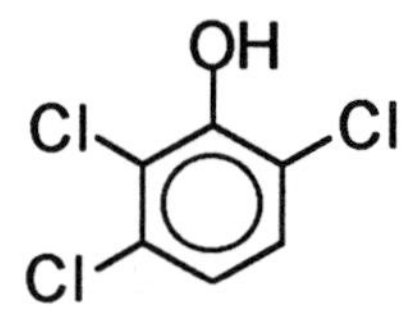

22. Which of the following compounds can be made from benzene by using two successive electrophilic substitution reactions?

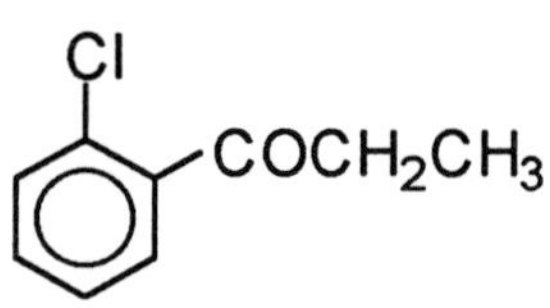
I

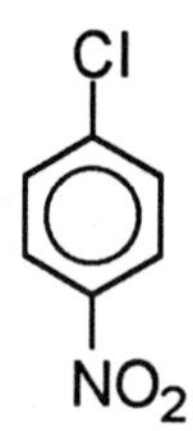
II

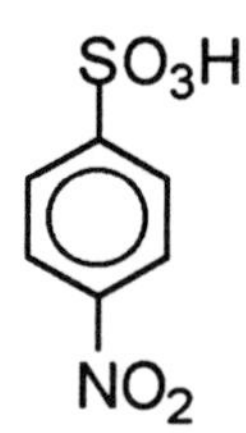
III

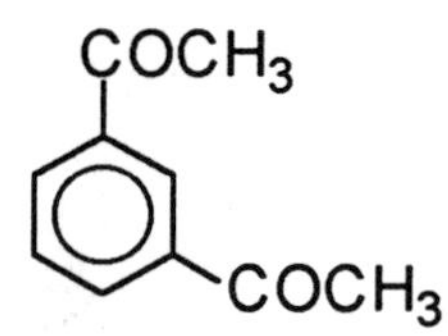
IV

a) I, II b) I, III, IV c) II, III, IV d) II, III

23. What is the major product from the following reaction?

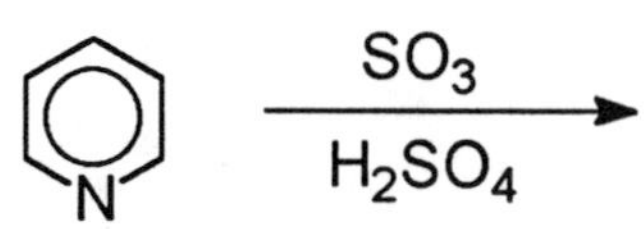

a) 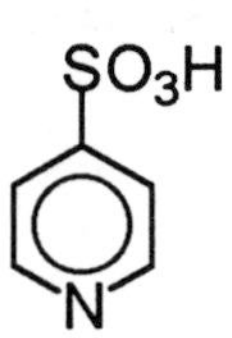

b) 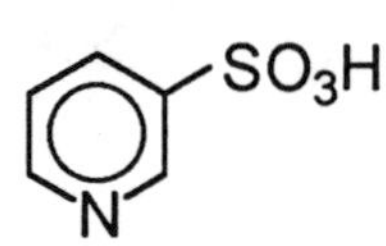

c) 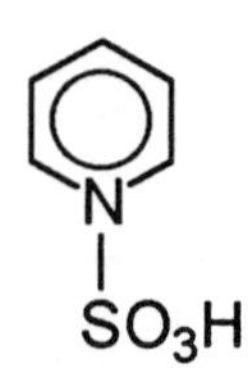

d) 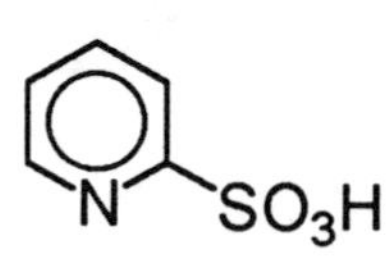

24. What is the major product from the following reaction?

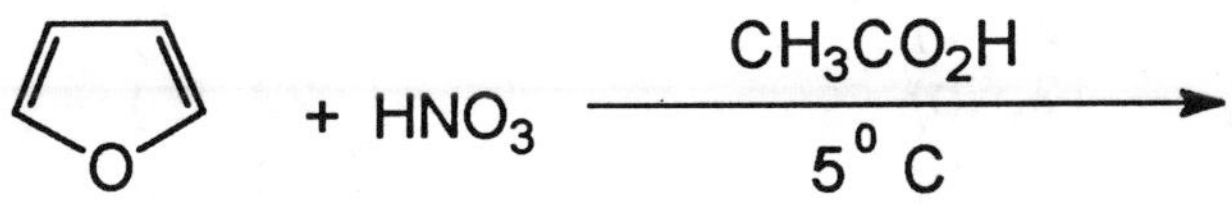

a) 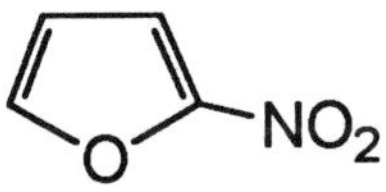 b) 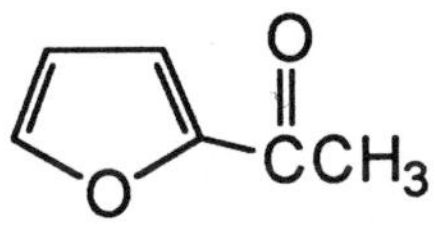 c) 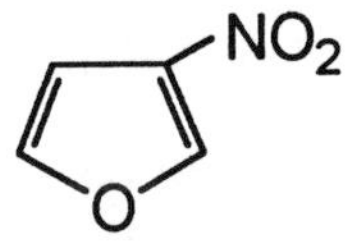 d) 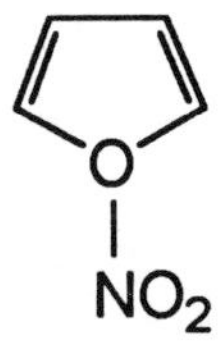

25. What is the major product from the following series of reactions?

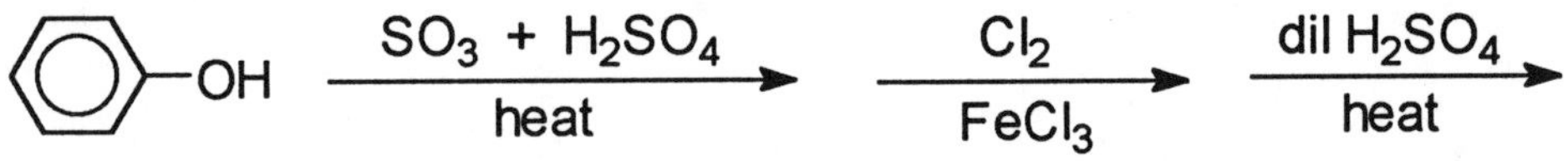

a) 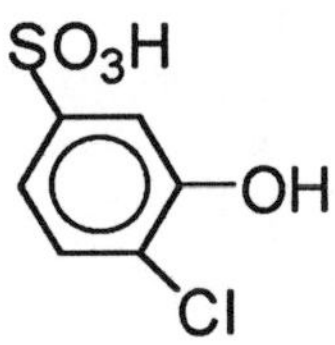 b) 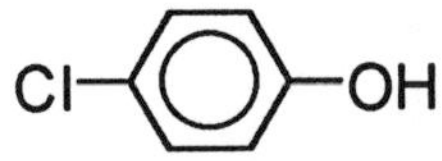

c) 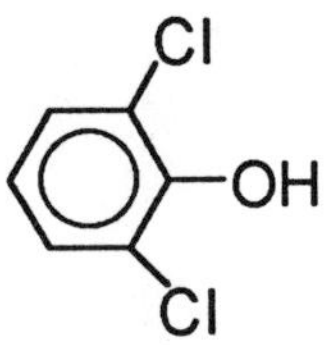 d) 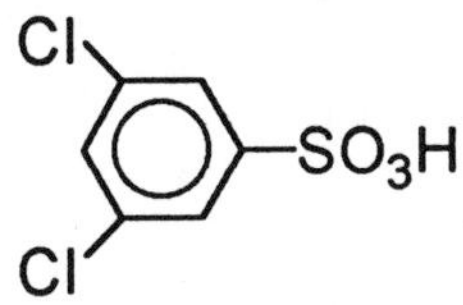

Answers

1. c)
2. b)
3. c)
4. a)
5. a)
6. b)
7. d)
8. b)
9. b)
10. b)
11. b)
12. c)
13. b)
14. d)
15. c)
16. d)
17. a)
18. a)
19. a)
20. a)
21. c)
22. a)
23. b)
24. a)
25. c)

Chapter 17 Aldehydes and Ketones

1. Which of the following compounds are named correctly?

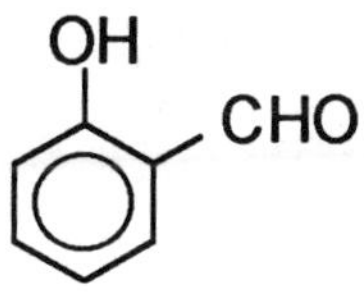

o-hydroxybenzaldehyde
I

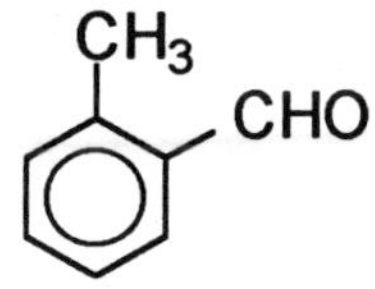

2-methylbenzaldehyde
II

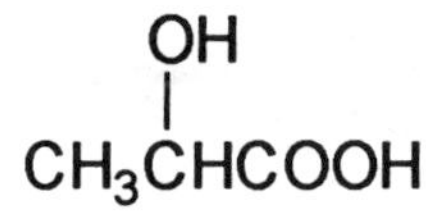

lactic acid
III

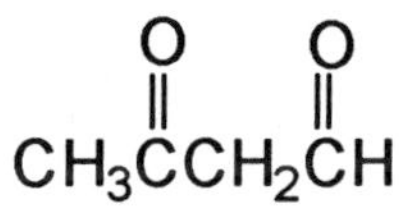

1,3-butanedione
IV

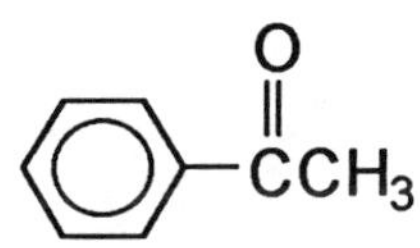

benzophenone
V

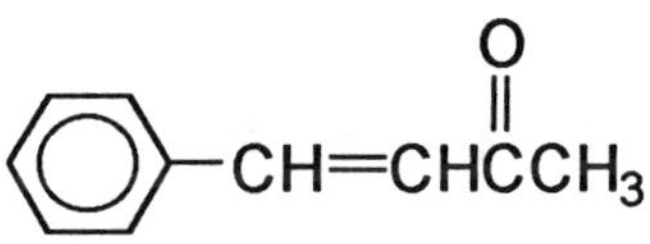

cinnamaldehyde
VI

a) I, II, III b) IV, V, VI c) II, IV, V d) III, IV, VI

2. What is the correct structure for 2-hydroxybenzophenone?

a) 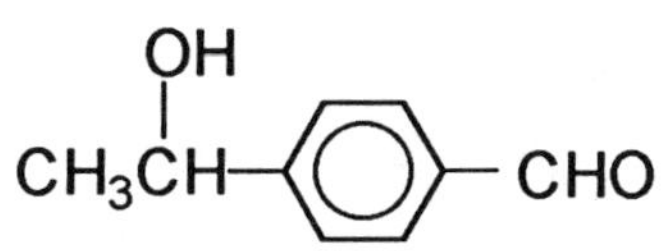

b) 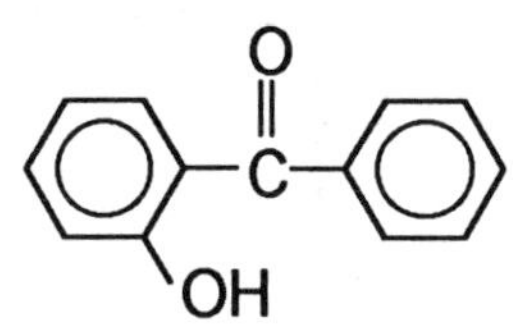

c) 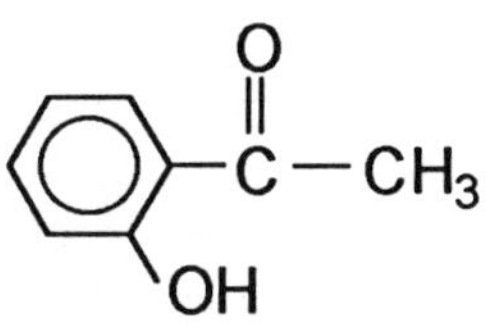

d) 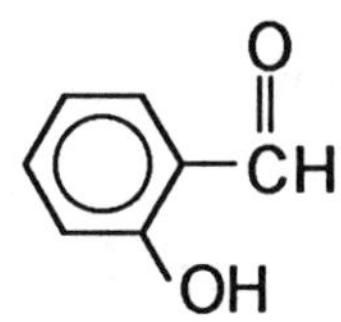

3. Which of the following compounds are acetals?

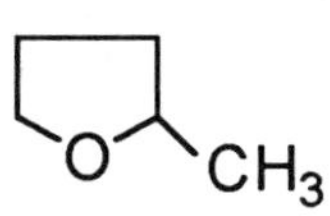

I

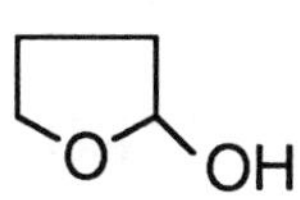

II

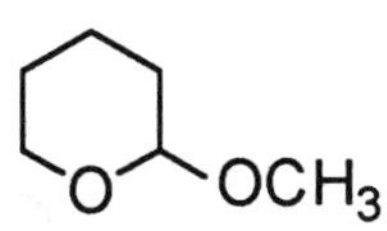

III

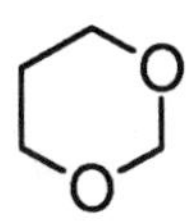

IV

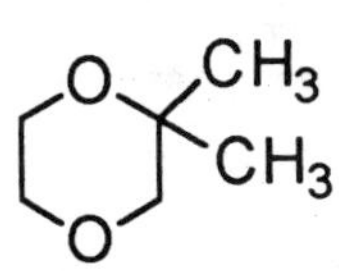

V

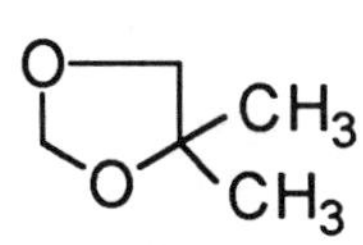

VI

a) I, II, III b) IV, V, VI c) III, IV, VI d) II, IV, V

4. Which of the following species are ylides?

I	II	III
$CH_3CH_2\overset{\oplus}{N}CH_3I^{\ominus}$	$Ph_3\overset{\oplus}{P}\overset{\ominus}{C}HCH{=}CH_2$	$Ph_3\overset{\oplus}{P}CH_3I^{\ominus}$

IV	V
$Ph_2\overset{\oplus}{S}\overset{\ominus}{C}HCH_3$	$PhN{=}CH_2$

a) I, II, III, IV b) I, V c) II, IV d) II, III, IV

5. A compound has a sharp, strong IR peak at ~1700 cm^{-1} and a weaker band at 2700 cm^{-1}. The proton NMR spectra has signals at 7.0 and 9.5 ppm in addition to other signals. What is the most likely structure for this compound?

a) 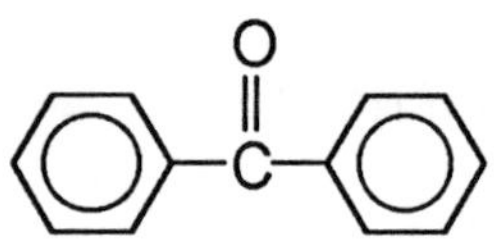

b) $CH_3CH_2\overset{O}{\overset{\|}{C}}CH{=}CH_2$

c) CH_3CH_2CHO

d) 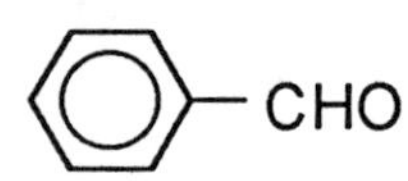

6. Which of the following reactions are good methods for preparing aldehydes?

I. $CH_3C{\equiv}CH + H_2O \xrightarrow[HgSO_4]{H_2SO_4}$

II. $CH_3CH_2CH_2OH \xrightarrow[CH_2Cl_2]{PCC}$

III. $(CH_3)_2CHCH{=}CHCH(CH_3)_2 \xrightarrow[OsO_4]{H_2O_2} \xrightarrow{Pb(OAc)_4}$

IV. $CH_3C{\equiv}CCH_3 \xrightarrow{BH_3\,/\,THF} \xrightarrow[NaOH]{H_2O_2}$

a) I, II b) II, III c) III, IV d) II, III, IV

7. Which of the following reactions does not lead to 3-methyl-3-hexanol?

a) 2-butanone + propyl magnesium bromide
b) 3-hexanone + methyl magnesium bromide
c) 2-pentanone + ethyl magnesium bromide
d) 3-pentanone + ethyl magnesium bromide

8. Which reagents are used for the following transformation?

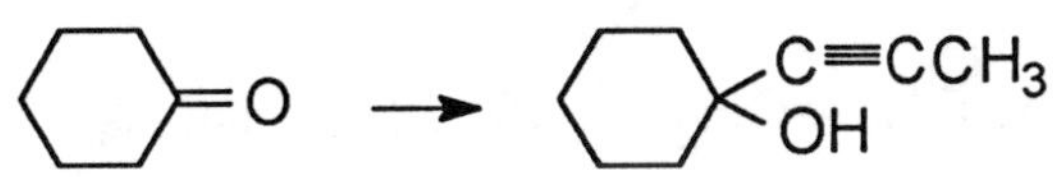

a) $\xrightarrow{CH_3C\equiv CH\ +\ H_2SO_4\ +\ H_2O}$

b) $\xrightarrow{\overset{\oplus}{Na}\ \overset{\ominus}{C}\equiv CCH_3}\ \xrightarrow[H_2O]{HCl}$

c) $\xrightarrow{CH_3CH=CHMgBr}\ \xrightarrow[H_2O]{HCl}$

d) $\xrightarrow{HC\equiv \overset{\ominus}{C}\ \overset{\oplus}{Li}}\ \xrightarrow{CH_3I}$

9. Which of the following enol forms is the most stable?

a) 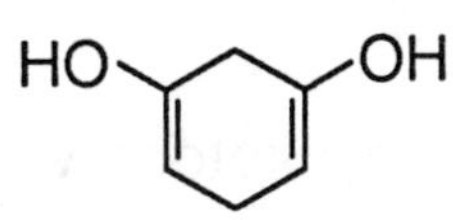

b) 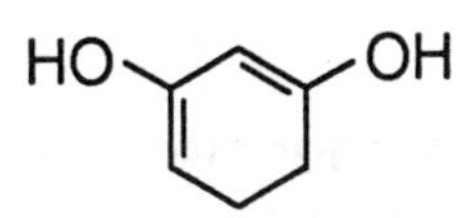

c) 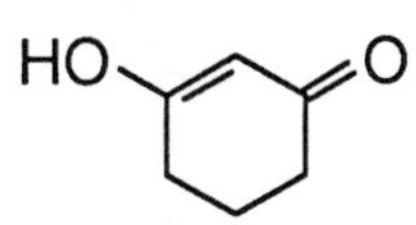

d) 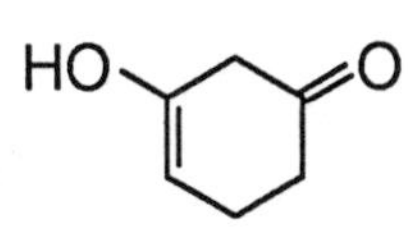

10. Which of the following reagents can distinguish between propanal and acetaldehyde?

a) hydrazine
b) the haloform reagents
c) silver nitrate in ammonia
d) sodium bisulfite

11. Which of the following compounds undergoes the fastest exchange of hydrogen for deuterium when treated with D_2O and a trace of ^{-}OD?

a) 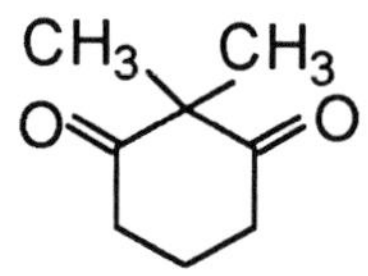 b) 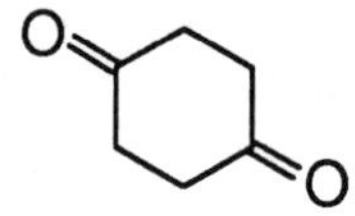 c) 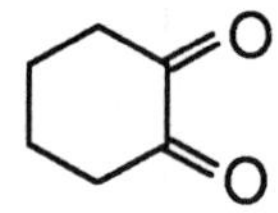 d) 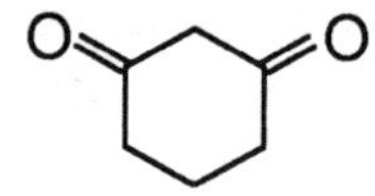

12. What is the product from a Wolff-Kishner reaction with acetaldehyde?

a) CH_3CH_3 b) $CH_2{=}CH_2$ c) $CH_3CH_2CH_3$ d) CH_3CH_2OH

13. What is the product from the reaction of cyclohexanone with hydroxyl amine?

a) 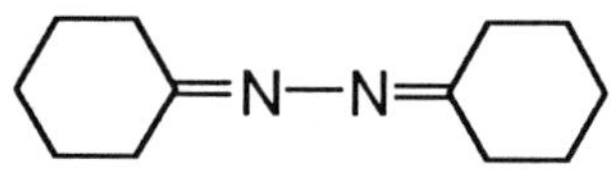 b) 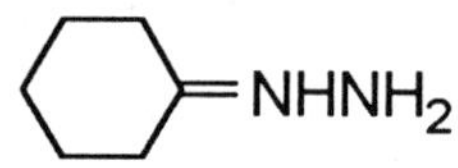

c) 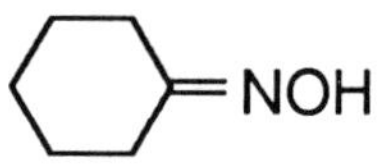 d) 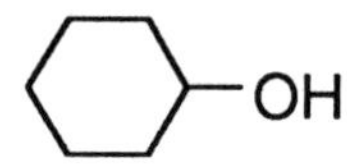

14. Which of the following reactions does not create a C - H bond?

a) Cannizzaro b) Wolff-Kishner c) Grignard d) Wittig

15. What is the product from the reaction of methyl vinyl ketone with the Wittig reagent, $CH_2 = P(C_6H_6)_3$?

a) 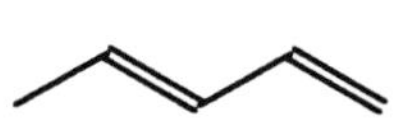 b) 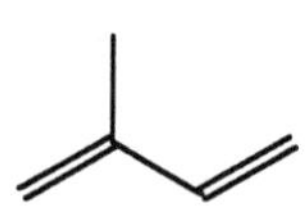 c) 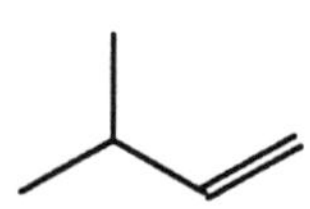 d) 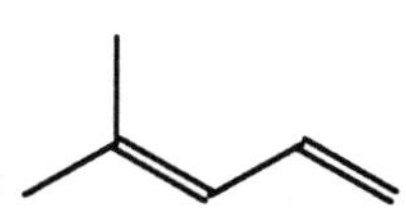

16. What is a major product from the following Cannizzaro reaction?

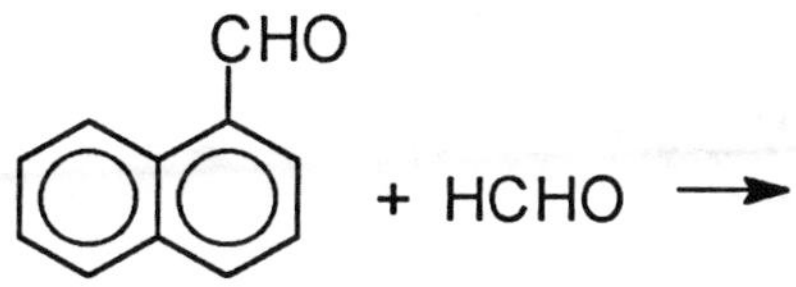

a) 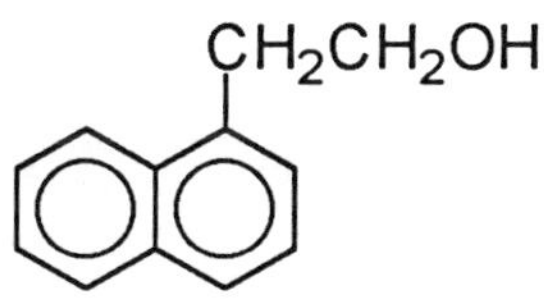

b) 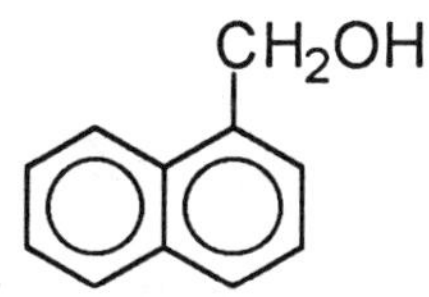

c) 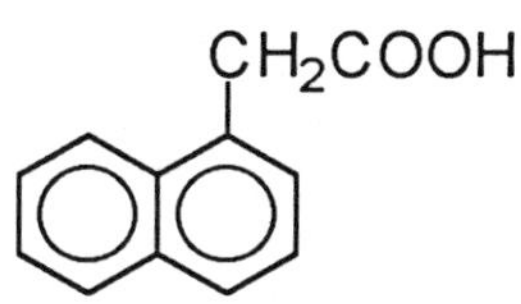

d) 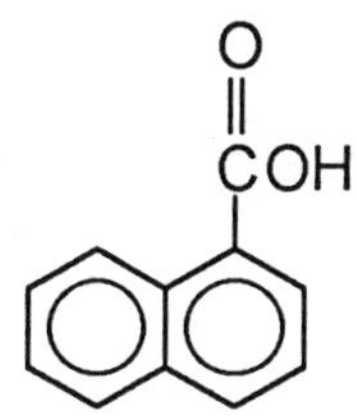

17. What is the major product from the Baeyer-Villiger oxidation of cyclohexanone?

a) $HOOC-(CH_2)_4-COOH$

b) 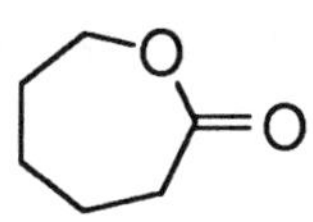

c) 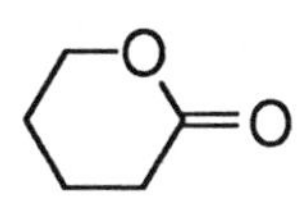

d) $HOOC-(CH_2)_3-COOH$

18. Which of the combinations of reactions is best for preparing the following compound?

$$HOCH_2C(CH_2OH)_3$$

a) aldol condensation / Cannizzaro

b) Grignard / Clemmensen

c) Wittig / Cannizzaro

d) Baeyer-Villiger

19. What is the product from the following series of reactions?

$$CH_3CH_2CHO \xrightarrow{^{\ominus}OH} \xrightarrow[\text{heat}]{H_3O^{\oplus}} \xrightarrow[\text{HCl}]{\text{Zn(Hg)}} \xrightarrow[\text{Pt}]{H_2}$$

a) $CH_3CH(CH_3)CH(CH_3)_2$ b) $CH_3CH_2CH_3$

c) $CH_3CH_2CH_2CH_2CH_2CH_3$ d) $CH_3CH_2CH_2CH(CH_3)_2$

20. What is the product from the following reaction?

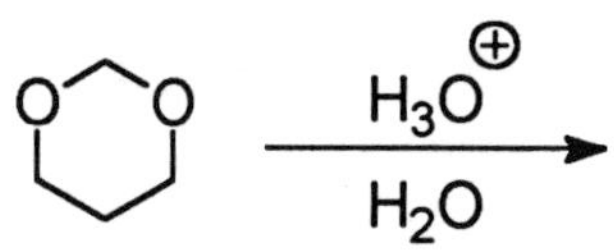

a) $CH_3CH_2CHO + CH_3OH$ b) $HOCH_2CH_2CH_2OH + C$

c) $CH_3C(=O)CH_2CH_2CH_2CHO$ d) $CH_2(CHO) + CH_3OH$

21. What is the major product from the following reaction?

$$HC(=O)(CH_2)_4C(=O)CH_3 + Na^{\oplus}\ ^{\ominus}OCH_2CH_3 \longrightarrow$$

a) 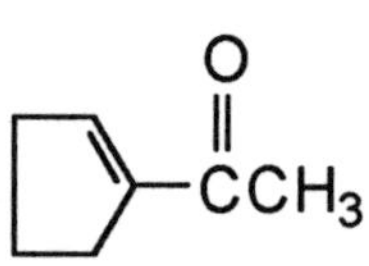 b) 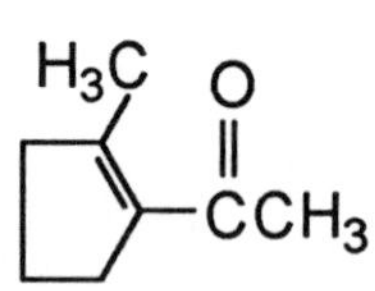 c) 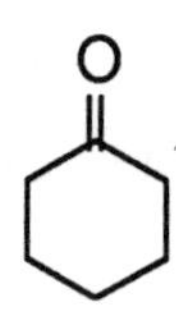 d) 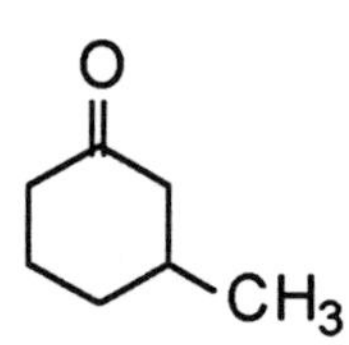

22. Which reaction condition cannot be used for the following conversion?

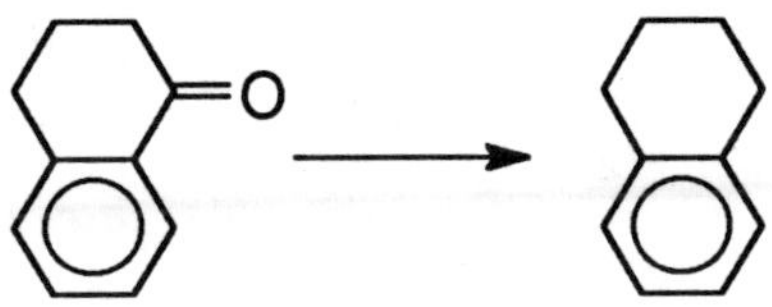

a) Zn(Hg) + HCl

b) NH_2NH_2 + KOH + heat

c) $\underset{SH}{\underset{|}{CH_2}}\underset{SH}{\underset{|}{CH_2}}$ + BF_3 then Raney - nickel

d) Platinum + hydrogen

23. What is the principal product from the following reaction?

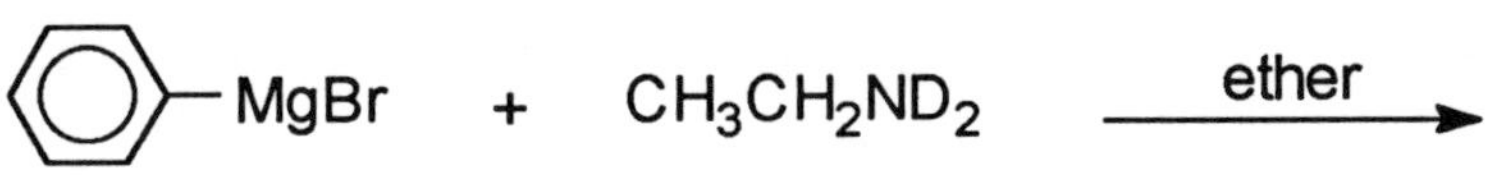

a) 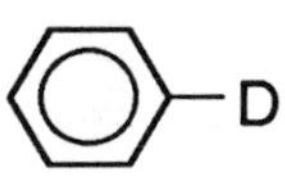 b) 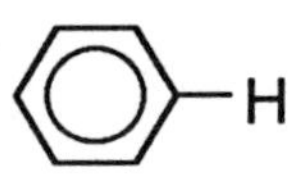 c) 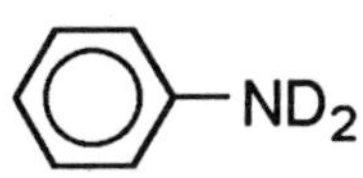 d) 

24. What is the principal product from the following reaction?

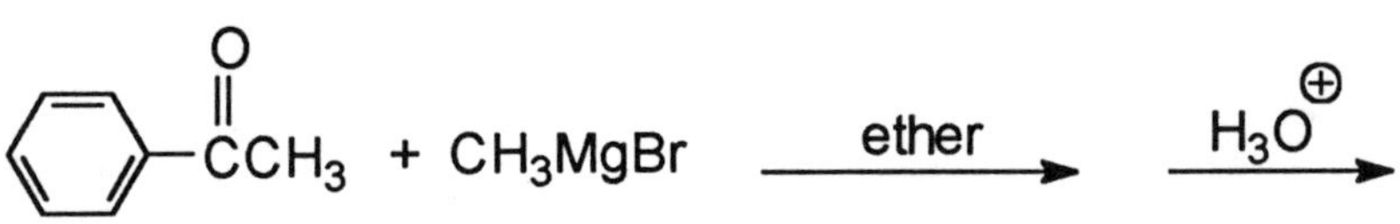

a) 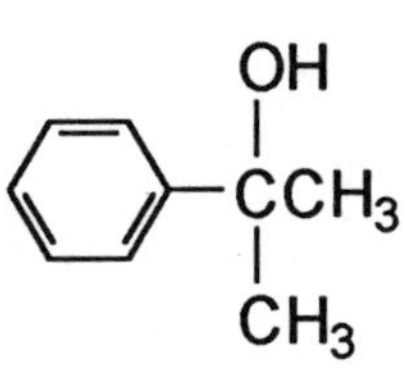

b) 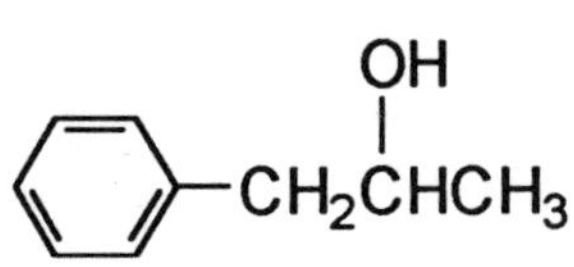

c) 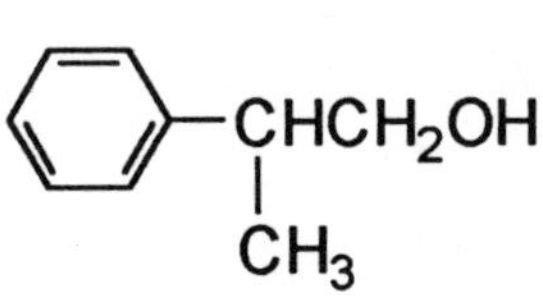

d) 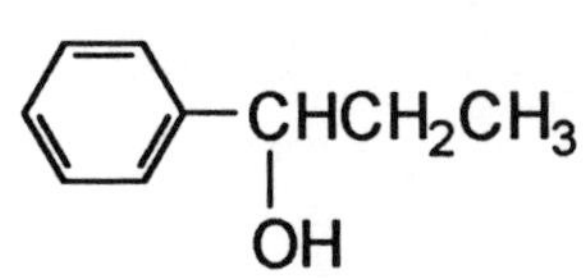

25. What is the best procedure for preparing pure ortho-deuteriotoluene from toluene?

a) $\xrightarrow[FeBr_3]{Br_2} \xrightarrow[ether]{Mg} \xrightarrow{D_2O}$

b) $\xrightarrow[H_2SO_4]{SO_3H} \xrightarrow[FeBr_3]{Br_2} \xrightarrow[ether]{Mg} \xrightarrow{D_2O} \xrightarrow[heat]{dil\ H_2SO_4}$

c) $\xrightarrow[H_2SO_4]{SO_3H} \xrightarrow[FeBr_3]{Br_2} \xrightarrow[heat]{dil\ H_2SO_4} \xrightarrow[ether]{Mg} \xrightarrow{D_2O}$

d) $\xrightarrow[FeBr_3]{Br_2} \xrightarrow[heat]{dil\ H_2SO_4}$

Answers

1. a)
2. b)
3. c)
4. c)
5. d)
6. b)
7. d)
8. b)
9. c)
10. b)
11. d)
12. a)
13. c)
14. c)
15. b)
16. b)
17. b)
18. a)
19. d)
20. b)
21. a)
22. d)
23. a)
24. a)
25. c)

Chapter 18 Carbohydrates

1. What is the correct structure for D-glyceraldehyde?

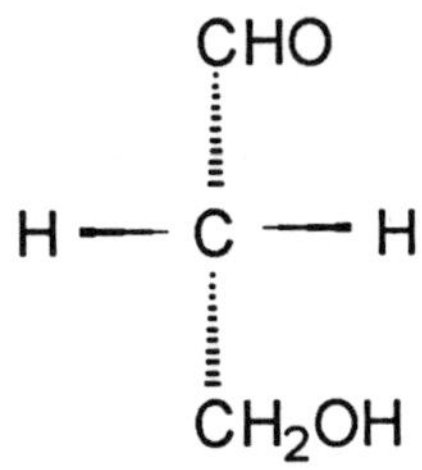
a)

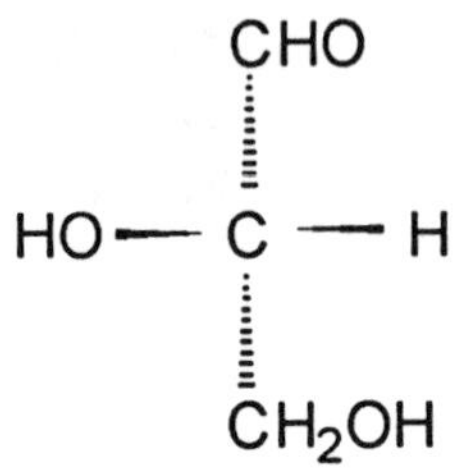
b)

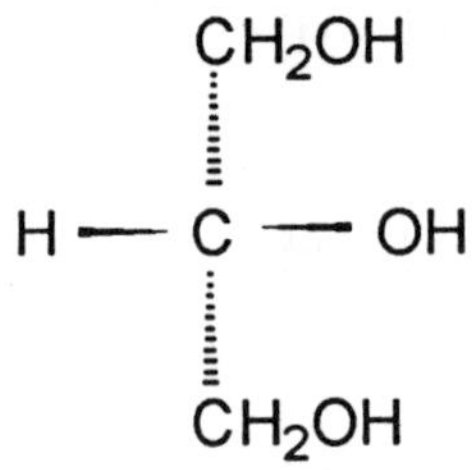
c)

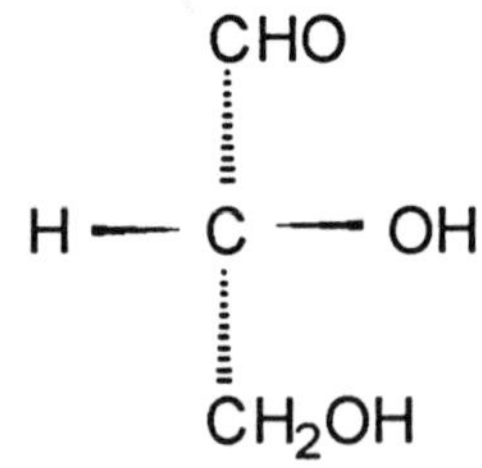
d)

2. Which of the following monosaccharides have the D configuration?

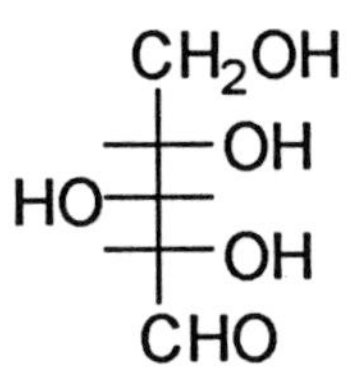
I

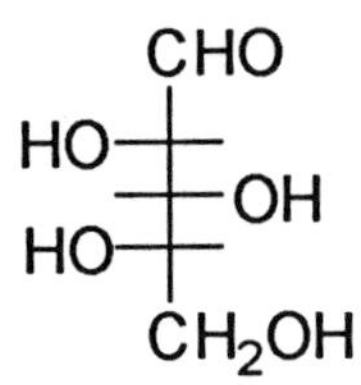
II

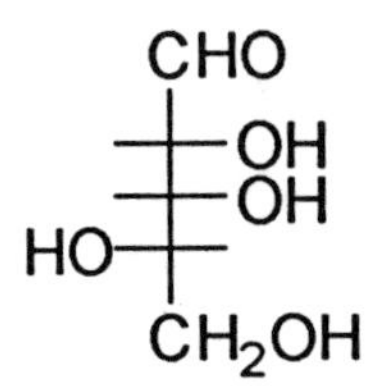
III

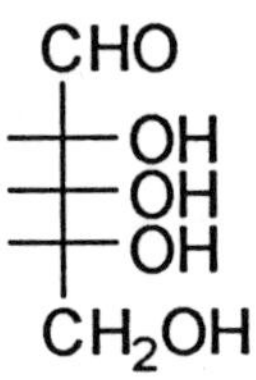
IV

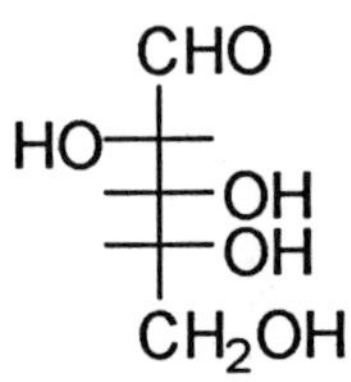
V

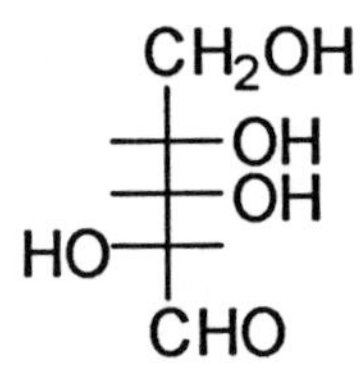
VI

a) I, IV, V b) II, III, VI c) IV, V, VI d) II, IV, V

3. What is the relationship between the following compounds?

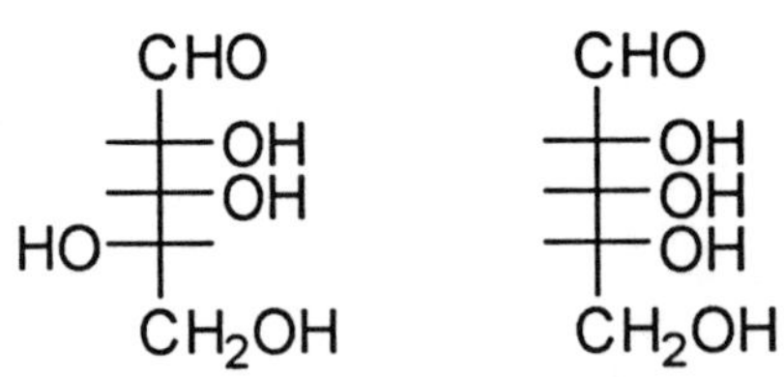

a) enantiomers b) anomers c) meso compounds d) epimers

4. What is the relationship between the following compounds?

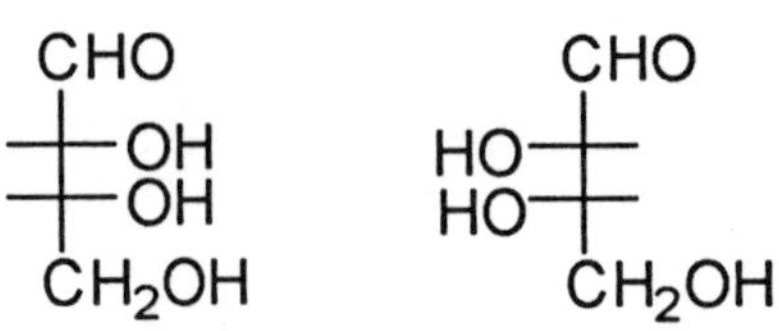

a) anomers b) epimers c) enantiomers d) equivalent forms

5. What is the relationship between the following compounds?

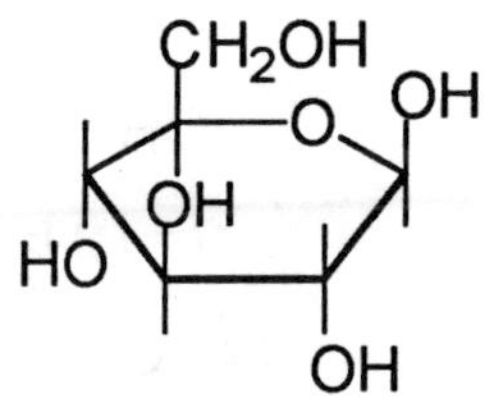

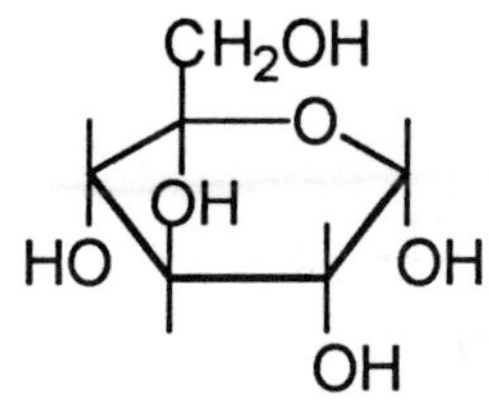

a) anomers b) epimers c) enantiomers d) equivalent forms

6. The following Fischer projection formula for L-idose corresponds to which Haworth projection formula?

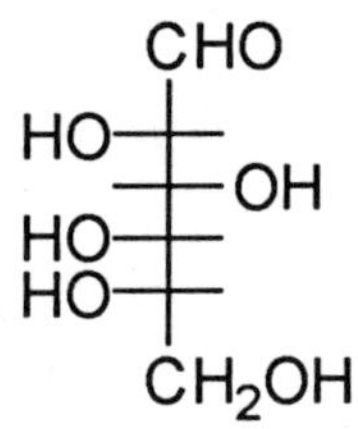

a) 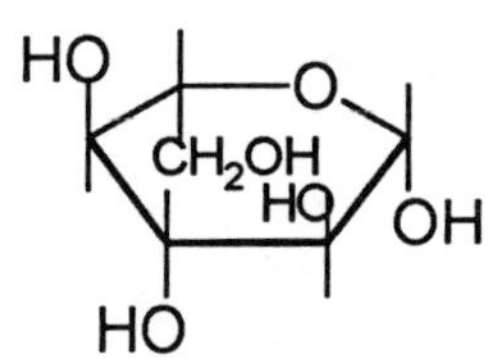

b) 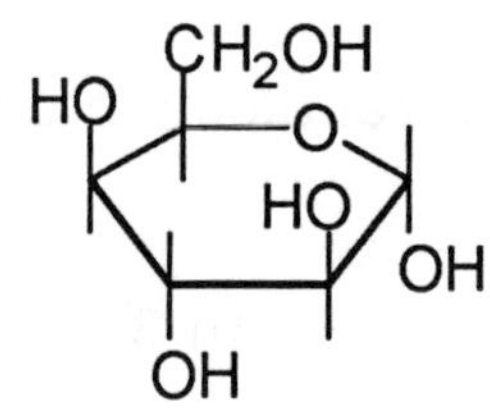

c) 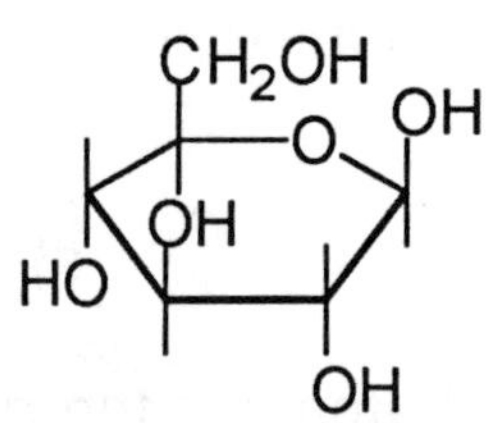

d) 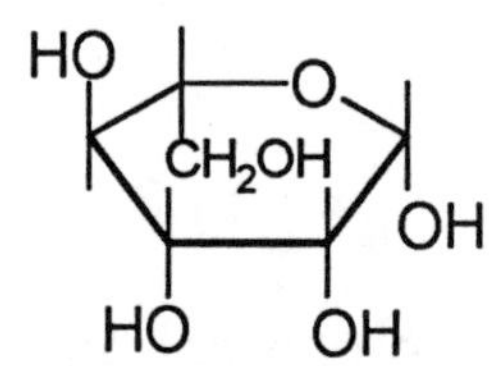

7. What is the correct structure for α-D-glucopyranose?

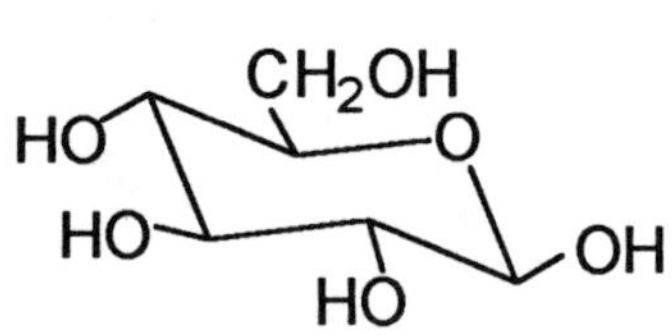

a)

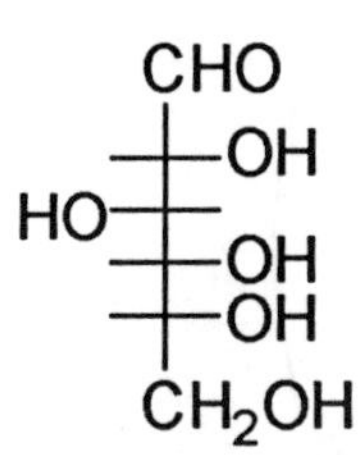

b)

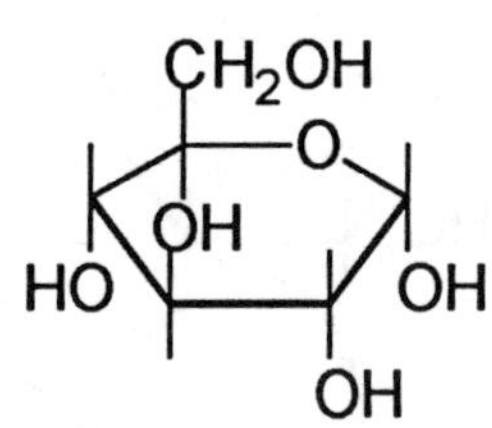

c)

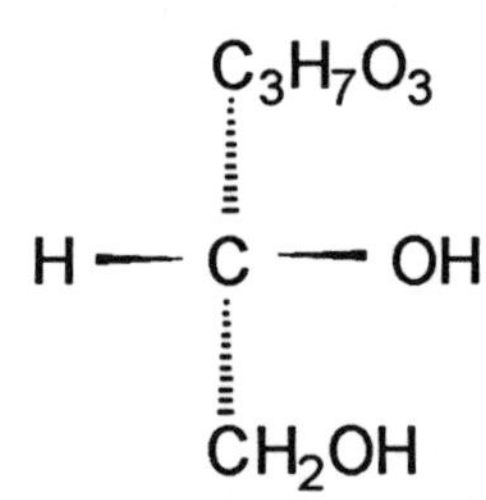

d)

8. What are the two major forms of D-fructose in equilibrium in aqueous solution?

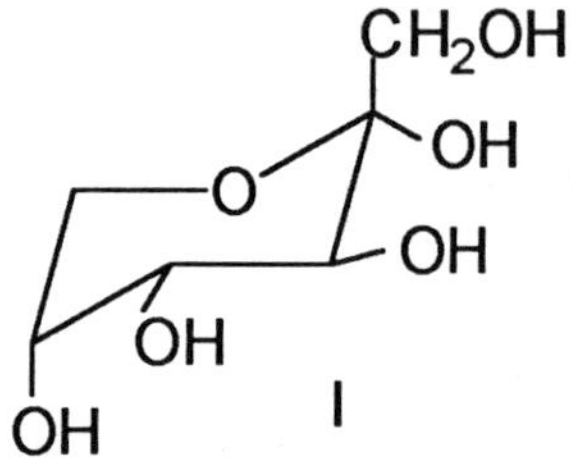

I

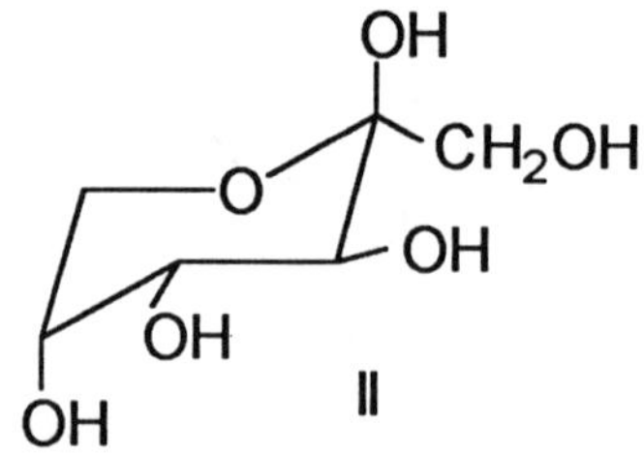

II

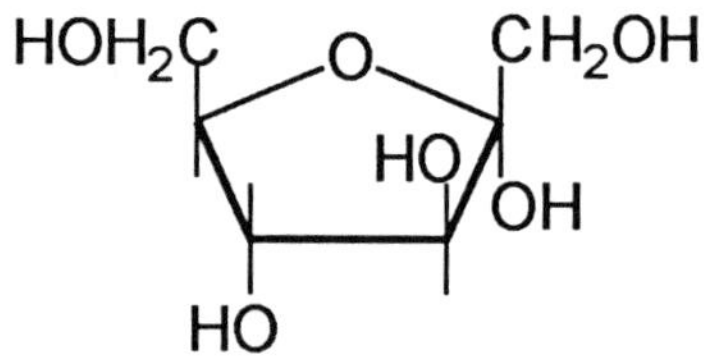

III

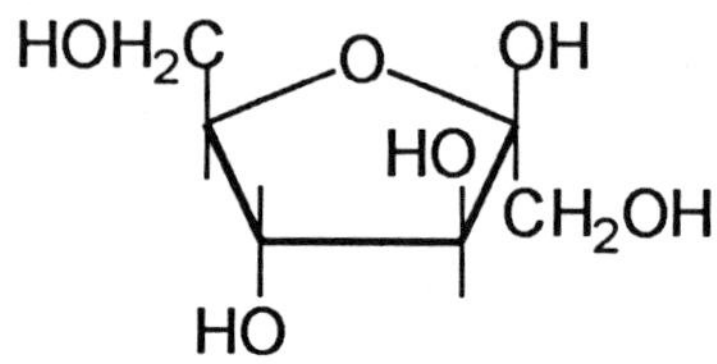

IV

a) I, II b) III, IV c) I, III d) II, IV

9. How many chiral atoms are there in a 2-ketohexose?

a) 2 b) 3 c) 8 d) 16

10. How many stereoisomers are possible for a D-aldohexose?

a) 2 b) 3 c) 8 d) 16

11. In the following compound, which of the labeled atoms is the anomeric carbon?

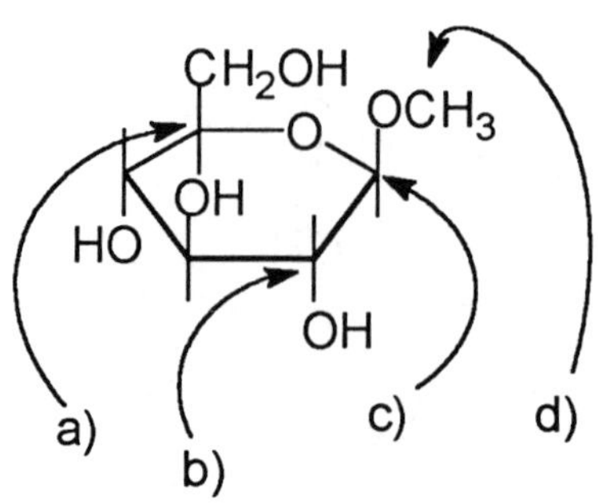

12. In the following compound, which of the labeled bonds is a glucoside?

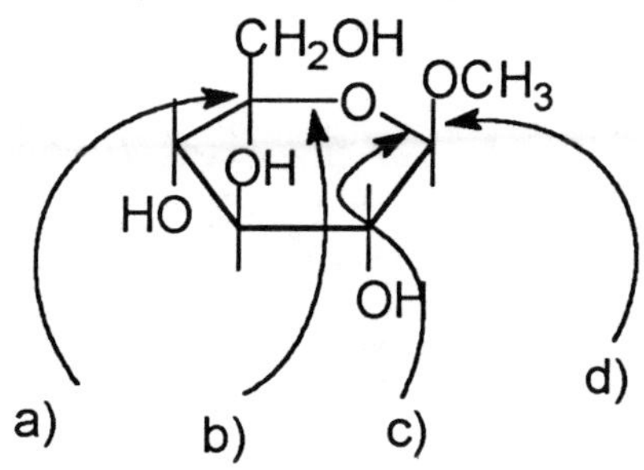

13. Which of the following compounds are glycosides?

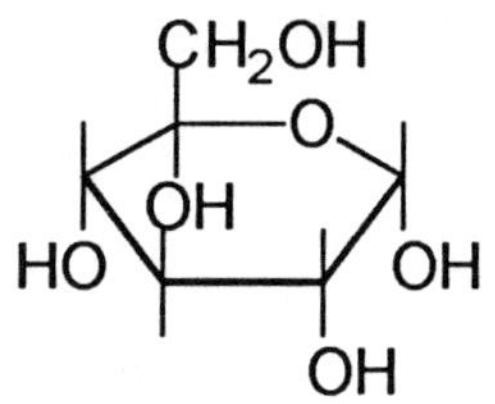

I

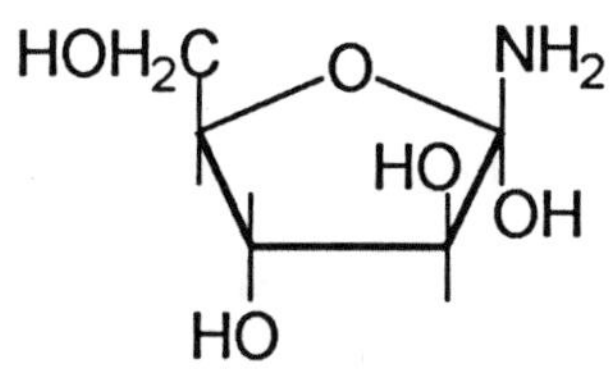

II

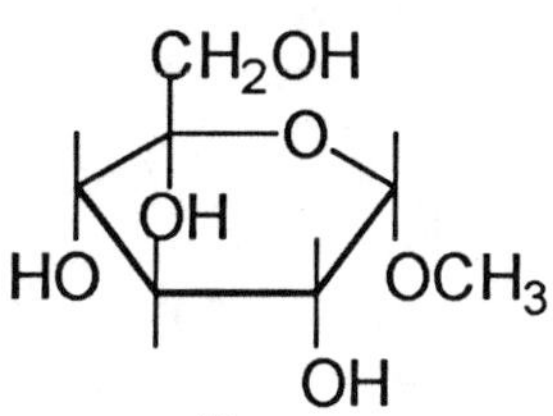

III

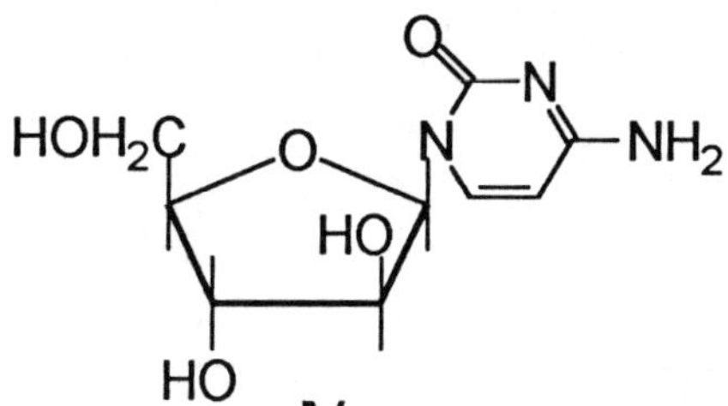

IV

a) I, II b) III, IV c) II, III d) I, IV

14. What reagent could be used to convert galactose to galactitol?

a) $NaBH_4$ in H_2O b) HNO_3, warm c) $AgNO_3$ in NH_3, H_2O d) H_5IO_6

15. What reagent could be used to convert galactose to galactaric acid?

a) $NaBH_4$ in H_2O b) HNO_3, warm c) $AgNO_3$ in NH_3, H_2O d) H_5IO_6

16. What reagent could be used to convert galactose to formic acid and formaldehyde?

a) $NaBH_4$ in H_2O b) HNO_3, warm c) $AgNO_3$ in NH_3, H_2O d) H_5IO_6

17. Which of the following are α sugars?

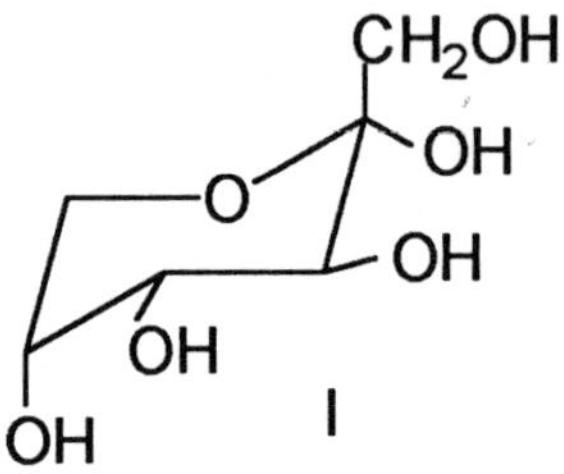
I

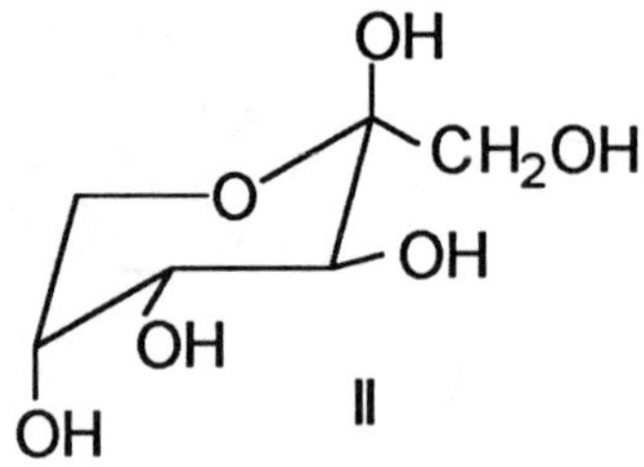
II

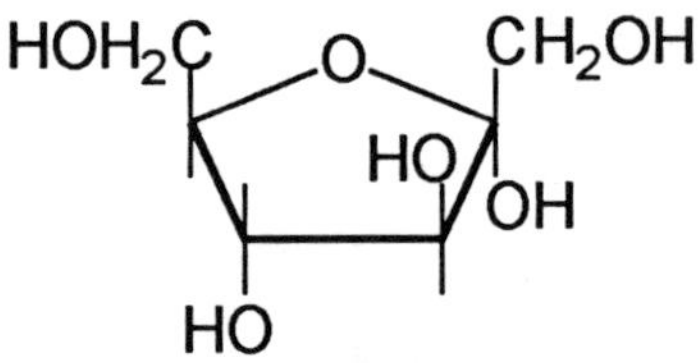
III

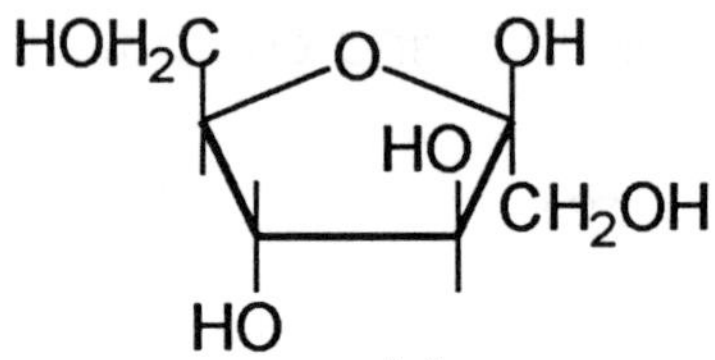
IV

a) I, II b) I, III c) II, IV d) II, IV

18. Two sugars give identical products when treated with sodium borohydride. Which of the following are possible structures for the sugars?

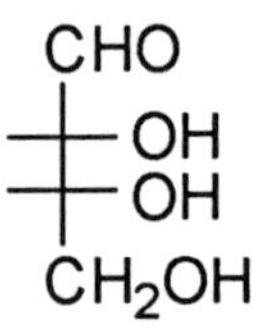
I

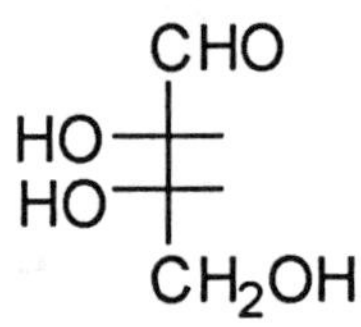
II

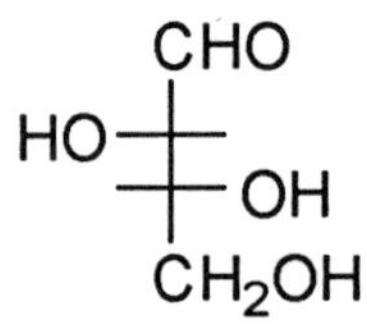
III

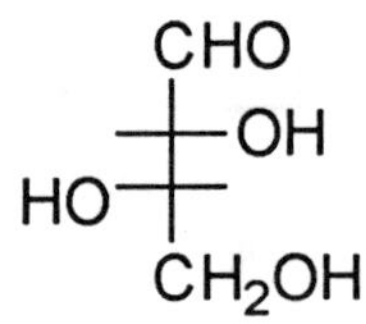
IV

a) I, II b) I, III c) III, IV d) II, IV

19. Which of the following sugars are disaccharides?

I. mannose II. maltose III. sucrose IV. lactose V. glucose

a) I, II, III b) II, III, IV c) II, IV, V d) I, III, IV

20. Cellulose, starch and glycogen are polysaccharides that contain which of the following sugars?

a) sucrose b) glucose c) fructose d) lactose

21. Which of the following statements about sucrose is NOT true?

a) The linkage between anomeric carbon atoms is 1,4'.
b) It is hydrolyzed to monosacchrides by invertases.
c) It cannot mutarotate nor reduce Tollens reagent.
d) It is a disacchride composed of glucose and fructose.

22. Which of the statements about the following sugar are true?

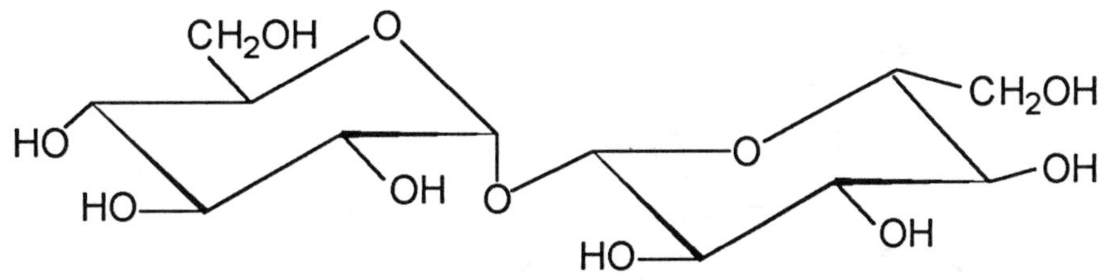

I. It is a reducing sugar.
II. It will undergo mutarotation.
III. The linkage is 1,1.
IV. It is composed of 2 units of D-glucose.

a) I, II b) II, III c) III, IV d) I, IV

23. The sugar in the following structure will react with X moles of HIO_4 to give Y moles of formaldehyde and Z moles of formic acid. What are the correct values for X, Y, Z?

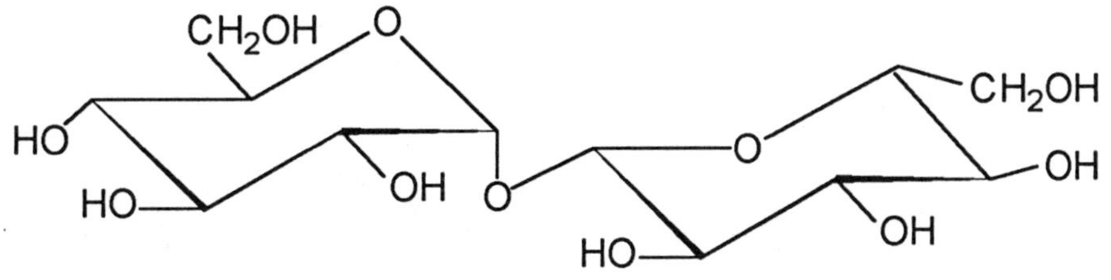

a) 10, 2, 10 b) 5, 1, 5 c) 10, 10, 2 d) 10, 2, 2

24. Which of the following reaction conditions will convert glucose to 2,3,4,6-tetra-O-methyl-D-glucose?

a) $\xrightarrow[HCl]{CH_3OH} \xrightarrow[NaOH]{(CH_3)_2SO_4}$

b) $\xrightarrow[HCl]{CH_3OH}$

c) $\xrightarrow[HCl]{CH_3OH} \xrightarrow[NaOH]{(CH_3)_2SO_4} \xrightarrow[HCl]{H_2O}$

d) $\xrightarrow[NaOH]{(CH_3)_2SO_4}$

25. Identify the most acidic proton in L-ascorbic acid.

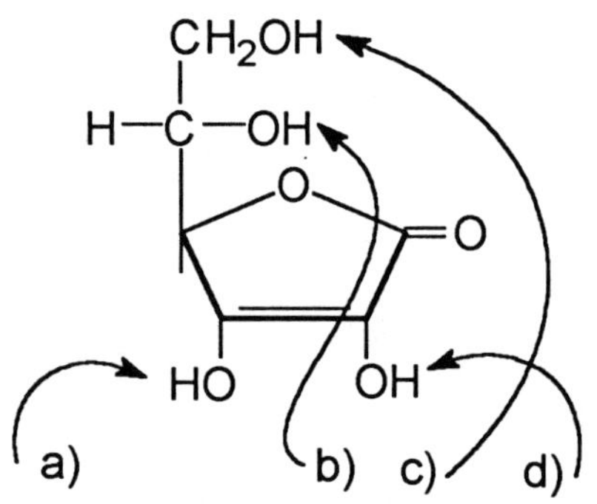

Answers

1. d)
2. c)
3. d)
4. c)
5. a)
6. a)
7. c)
8. d)
9. b)
10. c)
11. c)
12. d)
13. b)
14. a)
15. b)
16. d)
17. b)
18. a)
19. b)
20. b)
21. a)
22. c)
23. a)
24. c)
25. a)

Chapter 19 Carboxylic Acids

1. What is the correct name for the following structure?

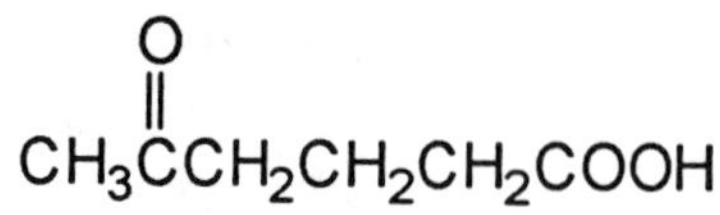

a) 2-oxo-hexanoic acid
b) 5-oxo-hexanoic acid
c) methyl butyroxo ketone
d) 4-ketopentanoic acid

2. What is the correct structure for Z-3-hexenedioic acid?

a) 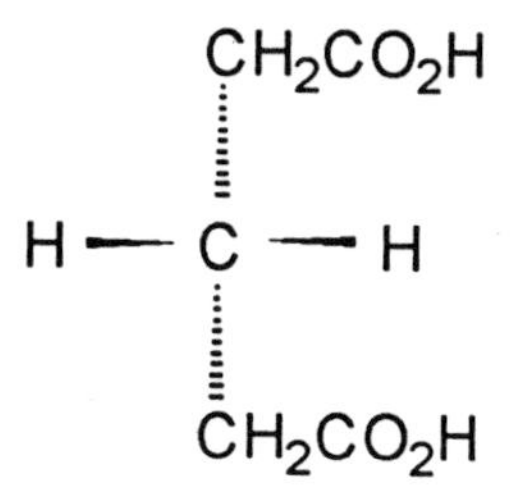

b) 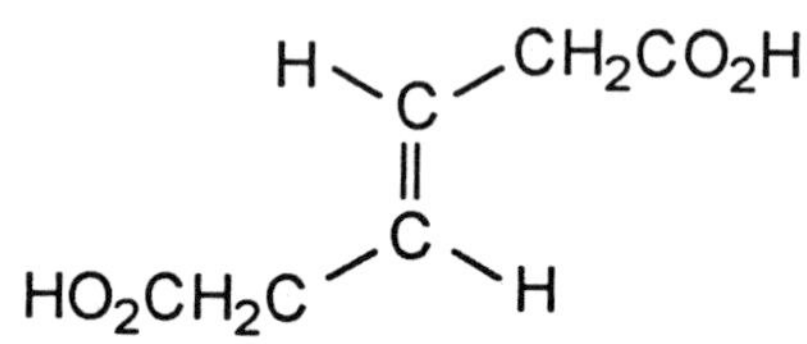

c) 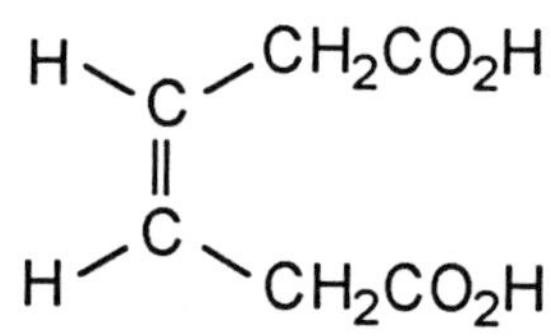

d) 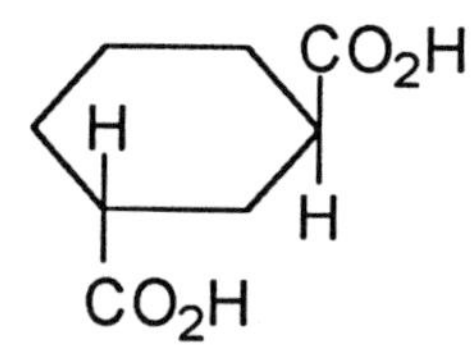

3. What is the correct name for the following structure?

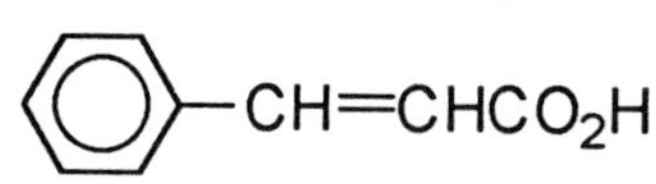

a) 3-phenylpropenoic acid
b) 3-carboxy-1-phenylethene
c) 1-phenylpropenoic acid
d) benzylacrylic acid

4. What is the correct structure for potassium acid tartrate?

a) 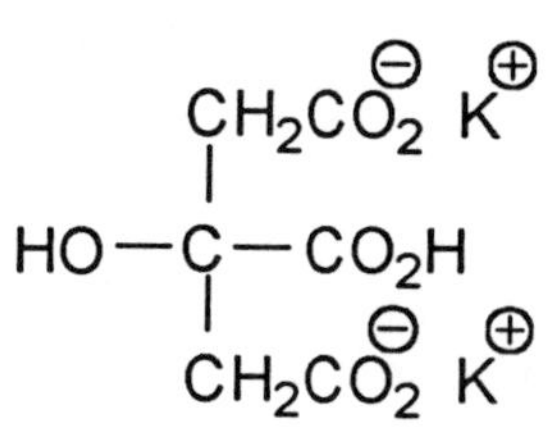

b) 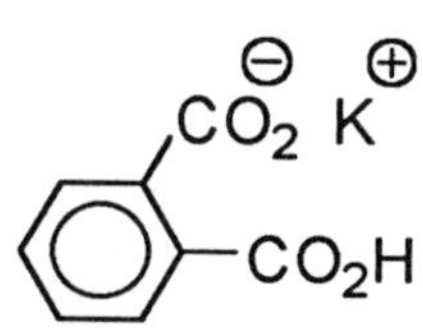

c) 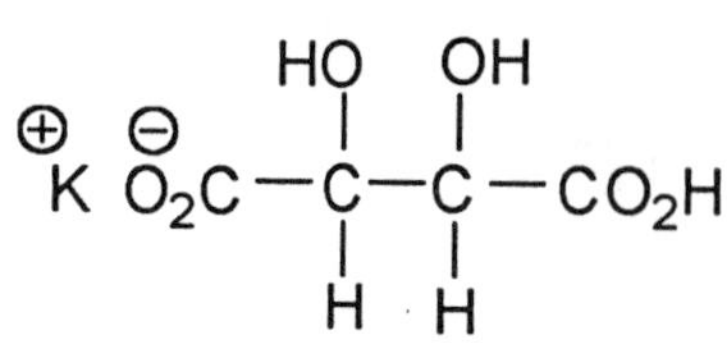

d) 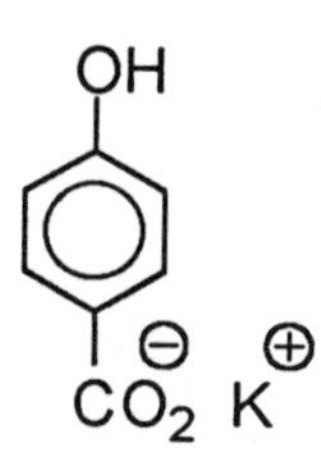

5. Which of the following structures are correctly named?

CH_2CO_2H	$CH_2{=}CHCO_2H$	$HO_2C{-}CH_2{-}CO_2H$	$HO_2C{-}CO_2H$
I	II	III	IV
formic acid	acrylic acid	malonic acid	succinic acid

a) I, II b) II, III c) III, IV d) II, IV

6. Which of the following acids is NOT soluble in water but is soluble in an aqueous solution of sodium hydroxide?

a) formic b) acetic c) butanoic d) benzoic

7. Arrange the following compounds in the order of increasing boiling point (lowest first).

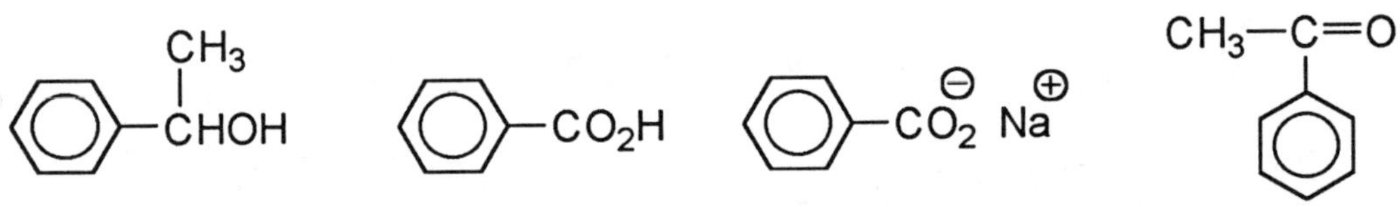

I II III IV

a) IV, II, I, III b) I, IV, III, II c) III, IV, I, II d) IV, I, II, III

8. What is the order of increasing acidity for the following compounds (lowest first)?

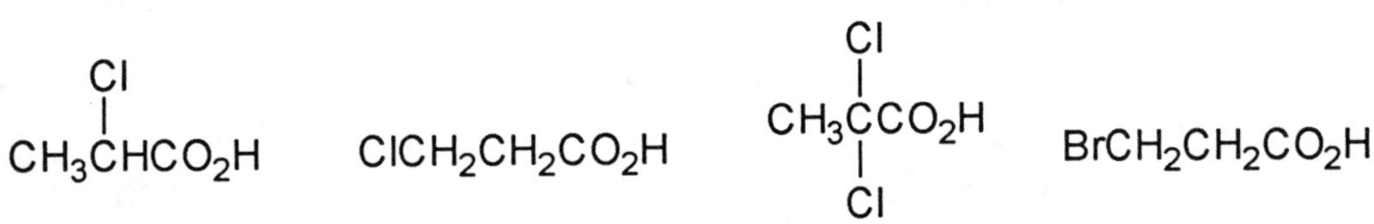

I II III IV

a) IV, I, II, III b) III, II, IV, I c) IV, II, I, III d) II, I, III, IV

9. What is the order of increasing acidity for the following compounds (lowest first)?

HCO_2H	$CH_3CH_2CO_2H$	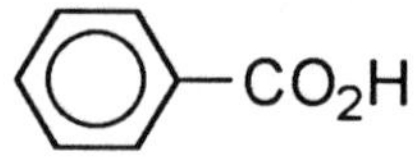	$ClCH_2CO_2H$
I	II	III	IV

a) III, II, I, IV b) I, II, IV, III c) II, III, IV, I d) II, I, III, IV

10. What is the order of increasing acidity for the following compounds (lowest first)?

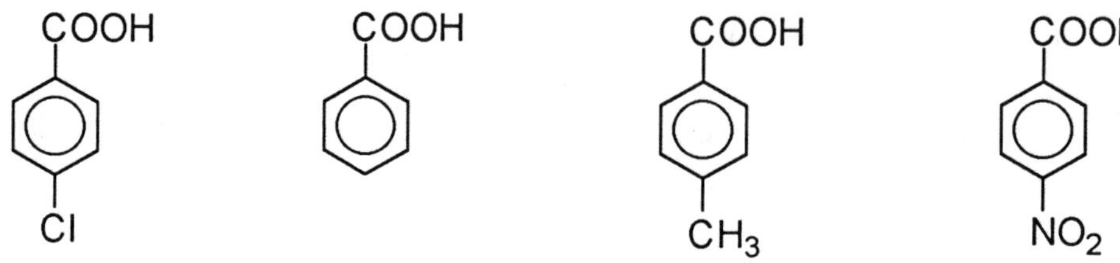

I II III IV

a) IV, I, III, I b) III, II, I, IV c) II, III, I, IV d) III, II, IV, I

11. Which of the following reactions proceed as written?

I. $CH_3CO_2H + NaOH \longrightarrow CH_3CO_2^{\ominus}\ Na^{\oplus} + H_2O$

II. $2\ CH_3CO_2H + Na_2CO_3 \longrightarrow 2\ CH_3CO_2^{\ominus}\ Na^{\oplus} + H_2CO_3$

III. $CH_3CO_2H + NaHSO_4 \longrightarrow CH_3CO_2^{\ominus}\ Na^{\oplus} + H_2SO_4$

IV. $CH_3CO_2H + C_6H_5CO_2^{\ominus}\ Na^{\oplus} \longrightarrow CH_3CO_2^{\ominus}\ Na^{\oplus} + C_6H_5CO_2H$

a) I, II b) III, IV c) I, III d) II, IV

12. Which of the following mixtures can be separated by treatment with aqueous NaOH?

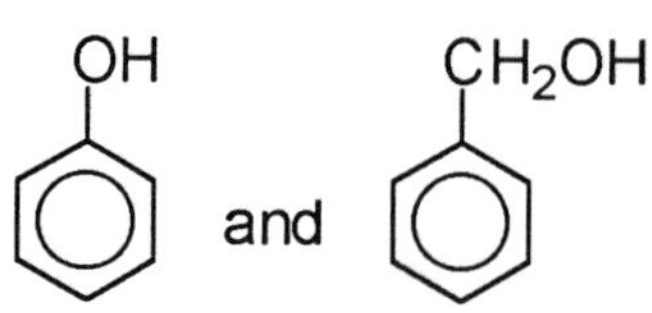

I

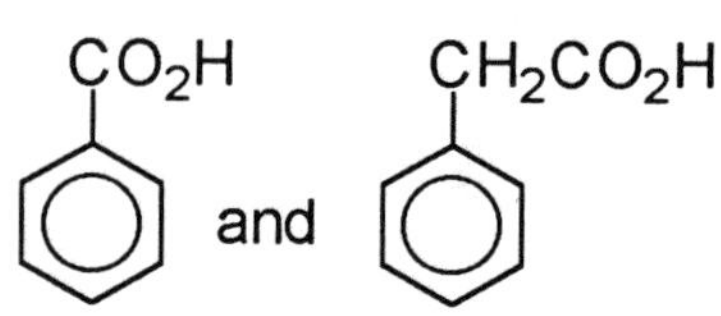

II

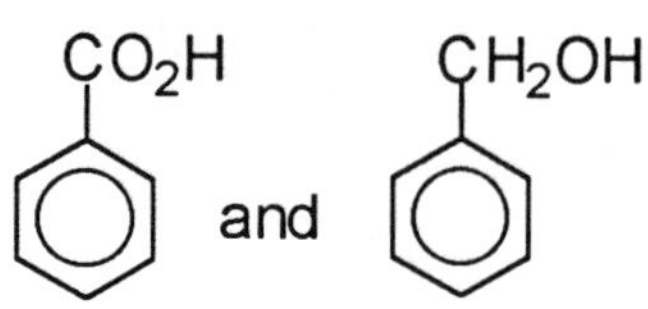

III

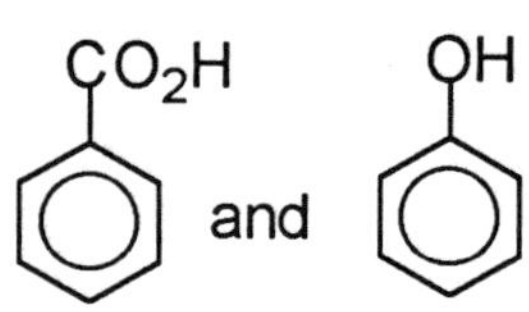

IV

a) I, II b) IIII, IV c) II, IV d) I, III

13. Which of the following mixtures can be separated by treatment with aqueous $NaHCO_3$?

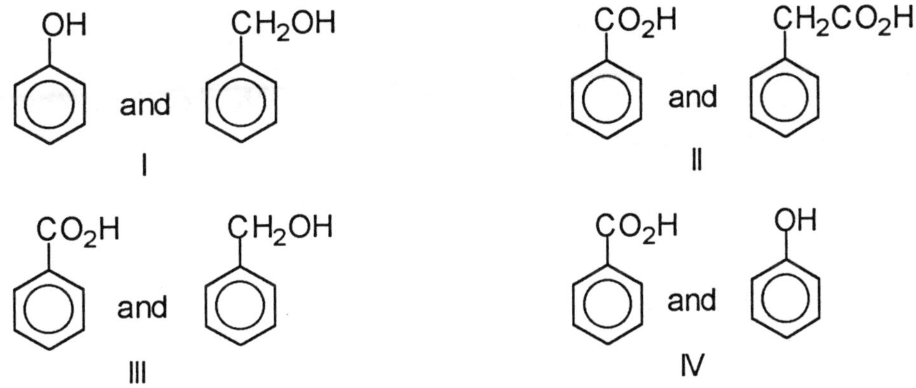

a) I, II b) IIII, IV c) II, IV d) I, III

14. Which carboxylic acid is prepared from the following reaction?

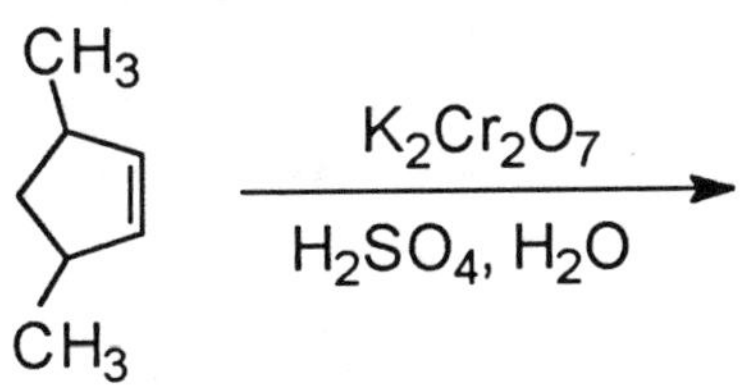

a) 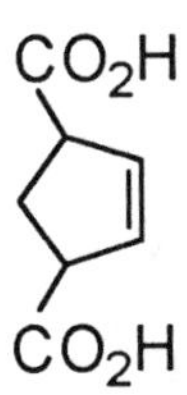 b) 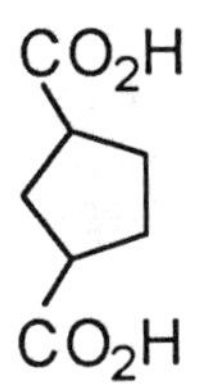 c) 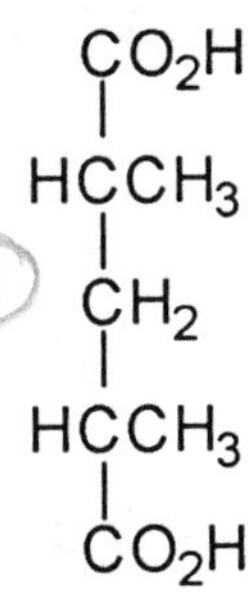 d) 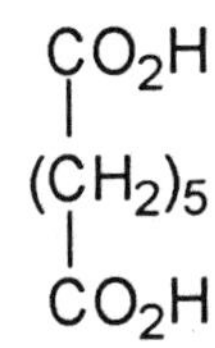

15. What are the best conditions for the following preparation?

a) $\xrightarrow[\text{cold}]{H_2CrO_4}$

b) $\xrightarrow{SOCl_2} \xrightarrow[\text{ether}]{Mg} \xrightarrow{CO_2} \xrightarrow[H_2O]{HCl}$

c) $\xrightarrow[H_2SO_4,\ H_2O]{K_2CrO_4}$

d) $\xrightarrow[\text{phase - transfer catalyst}]{KMnO_4,\ KOH,\ H_2O}$

16. Which acid is prepared by reacting acetylene and water with sulfuric acid and mercuric sulfate, followed by treatment with oxygen and cobalt (III) acetate?

a) citric acid b) succinic acid c) acetic acid d) acetoacetic acid

17. What is the product from the reaction of toluene with alkaline potassium permanganate, followed by reaction with ethanol and sulfuric acid?

a) 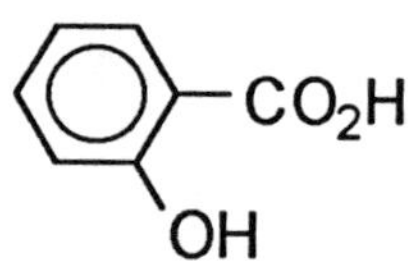

b) 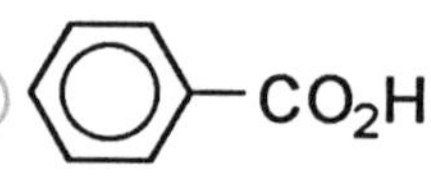

c) 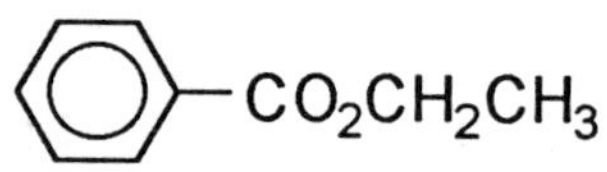

d) 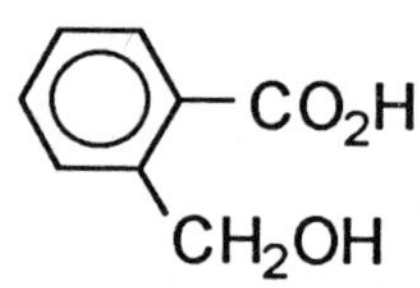

18. Which of the following reactions yield the same carboxylic acid?

I. 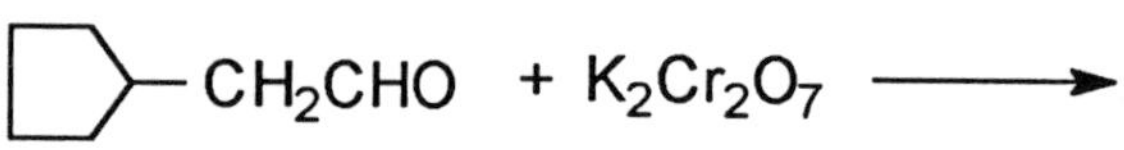

II. 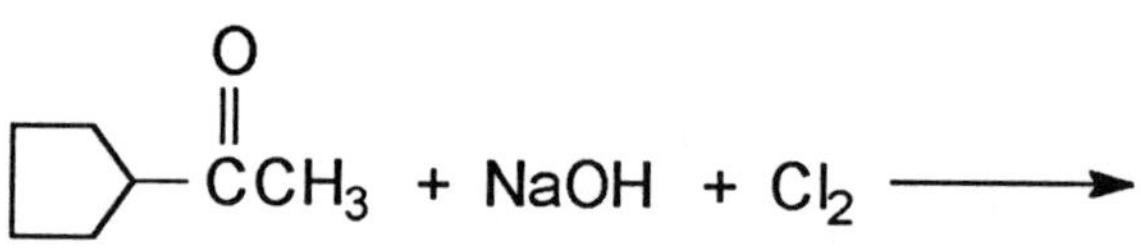

III. 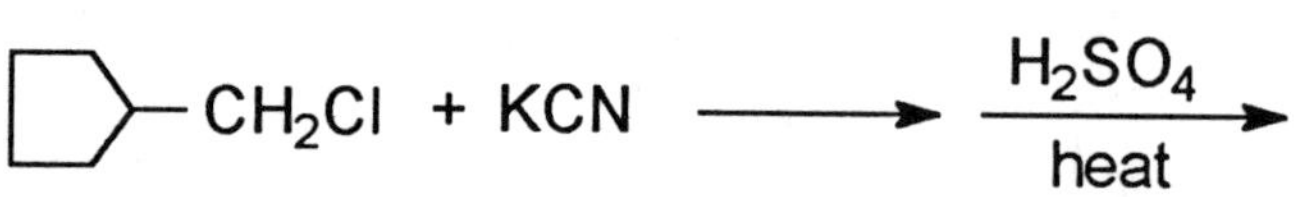

IV. 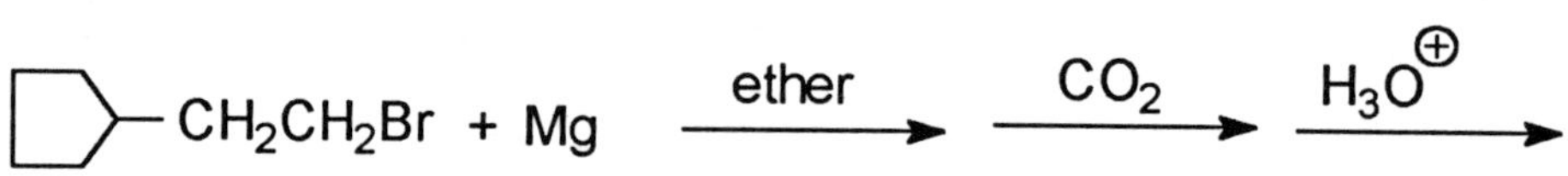

a) I, II b) II, III c) III, IV d) I, III

19. What is the product from the following reaction?

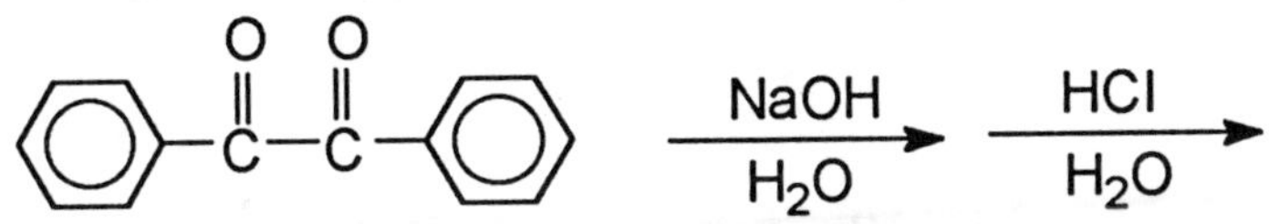

a) 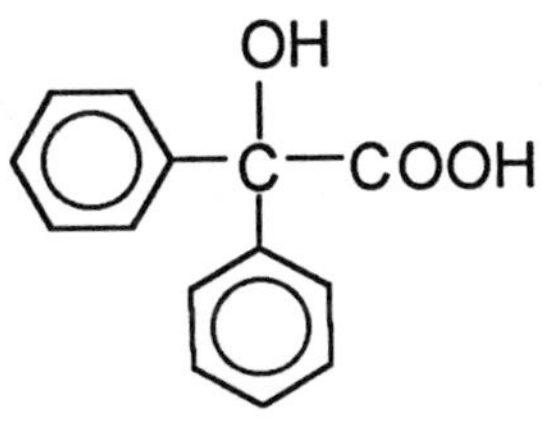

b) 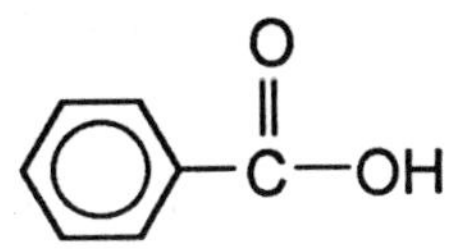

c) 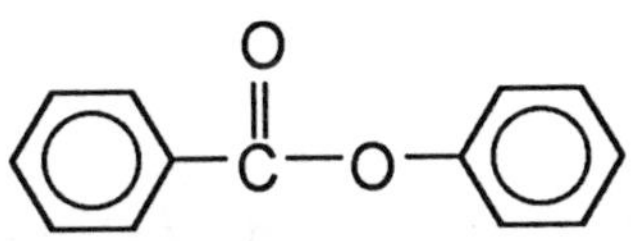

d) 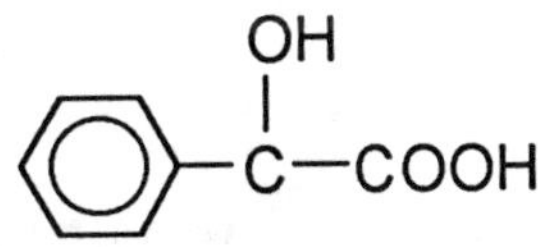

20. Which of the following compounds undergo thermal decarboxylation?

a) 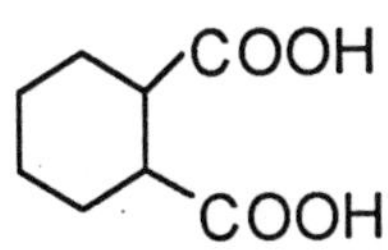

b) 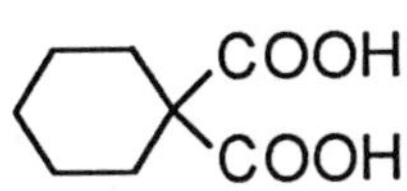

c) 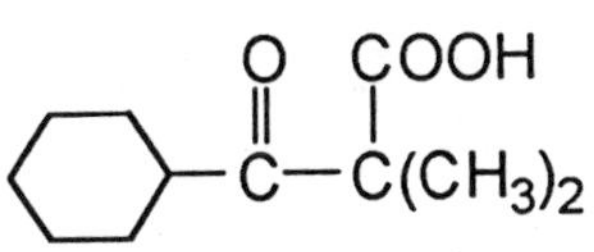

d) 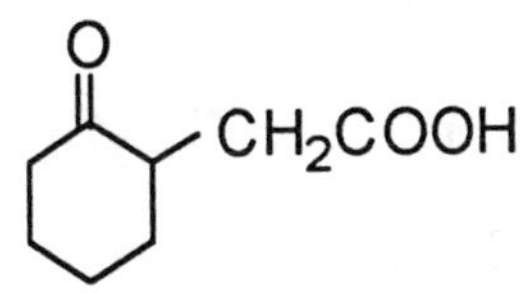

21. What is the product from the reaction of phenyl magnesium bromide and acetic acid?

a) 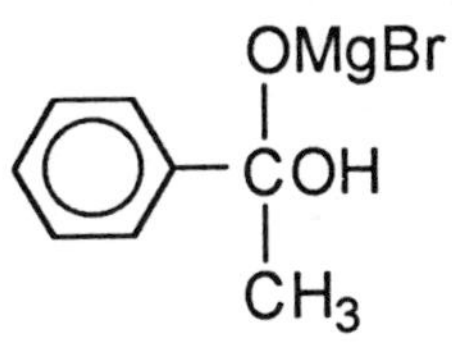

b) 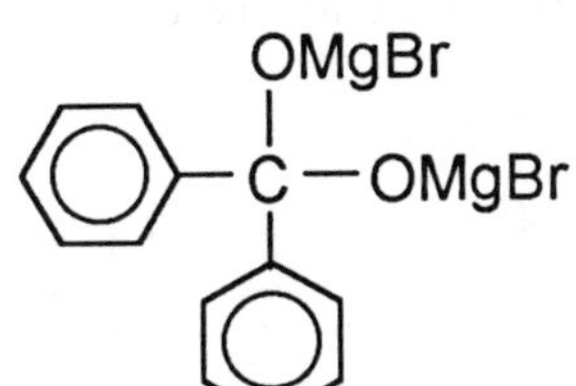

c) 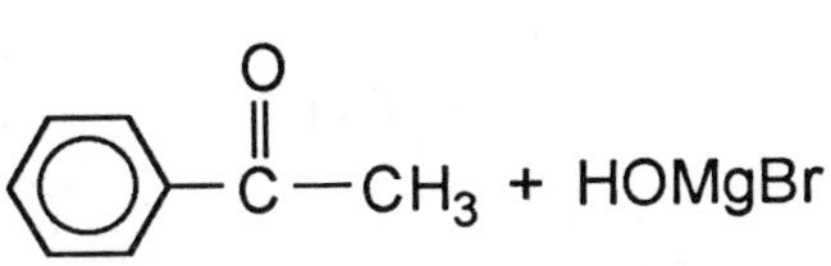

d) 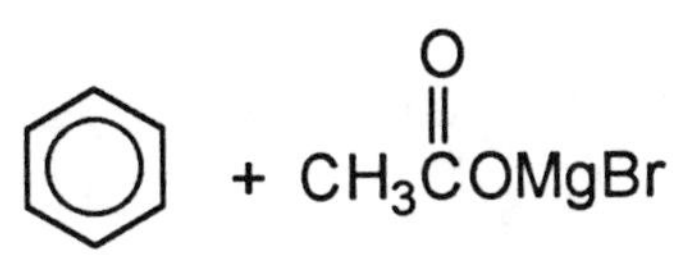

22. Phenol is treated with carbon dioxide in basic solution. The product from this reaction is reacted with thionyl chloride. What is the likely structure for the final product?

a) 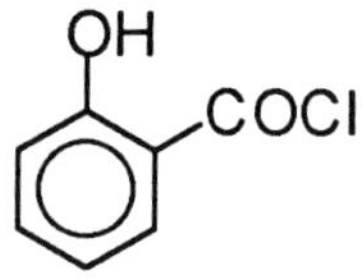 b) 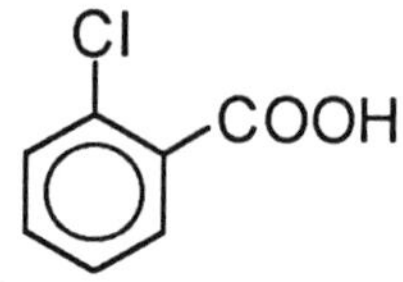 c) 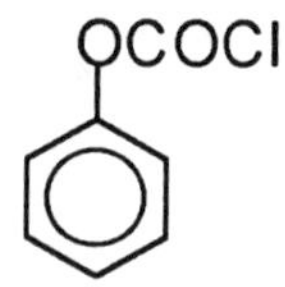 d) 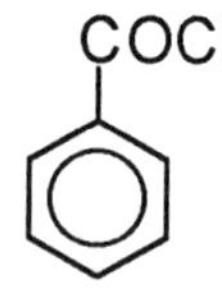

23. Which is the best reagent for the following transformation?

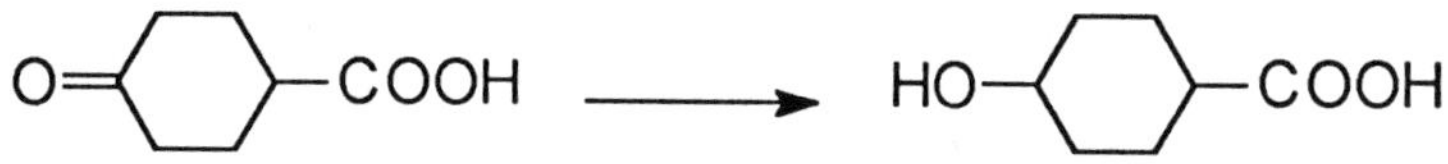

a) $LiAlH_4$ b) $NaBH_4$ c) H_2 + Pt d) pyridine • Cr_3O_4

24. Which of the conditions will give as a product the following compound?

$$CH_3CH_2CH_2O\overset{\overset{\displaystyle O}{\|}}{C}CH_3$$

a) $CH_3OH + CH_3CH_2COOH + HCl$

b) $CH_3COOH + CH_3CH_2COOH + HCl$

c) $CH_3COOH + CH_3CH_2CH_2OH + HCl$

d) $CH_3CH_2CH_2ONa + CH_3COOH$

25. What is the most likely product from the following series of reactions?

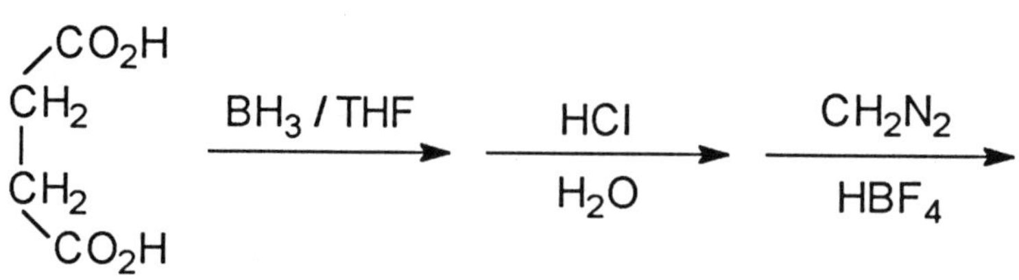

a) 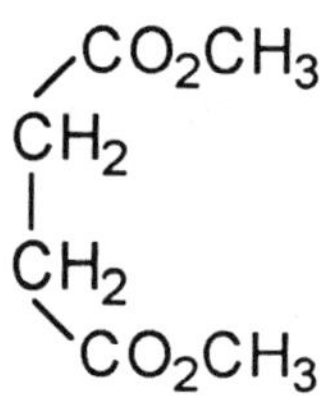 b) 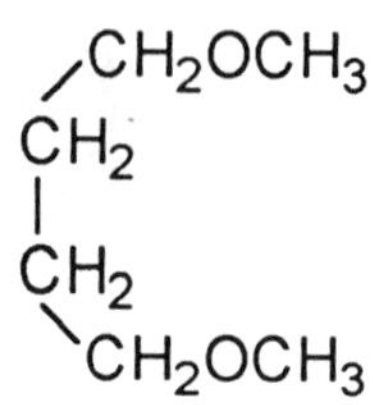 c) 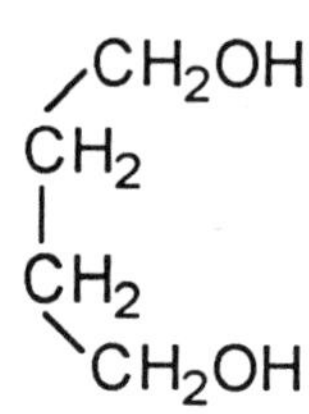 d) 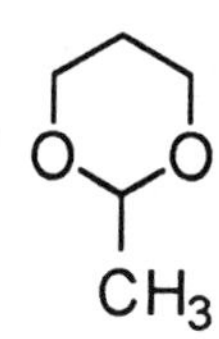

Answers

1. b)
2. c)
3. a)
4. c)
5. b)
6. d)
7. d)
8. c)
9. a)
10. b)
11. a)
12. d)
13. b)
14. c)
15. b)
16. c)
17. c)
18. d)
19. a)
20. b)
21. d)
22. a)
23. b)
24. c)
25. b)

Chapter 20 Functional Derivatives of Carboxylic Acids

1. What is the correct structure for phenylbenzoate?

a) 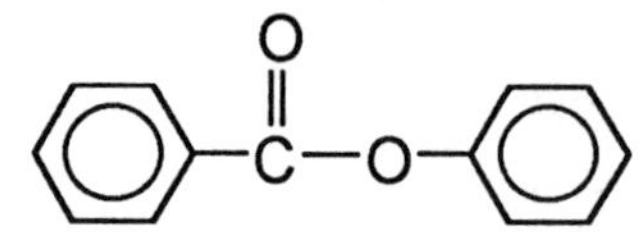

b) 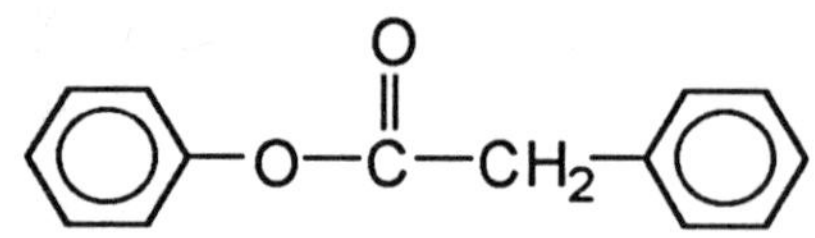

c) 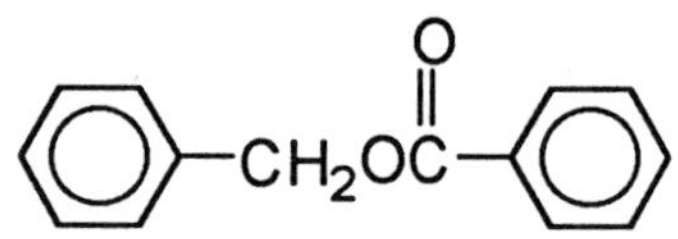

d) 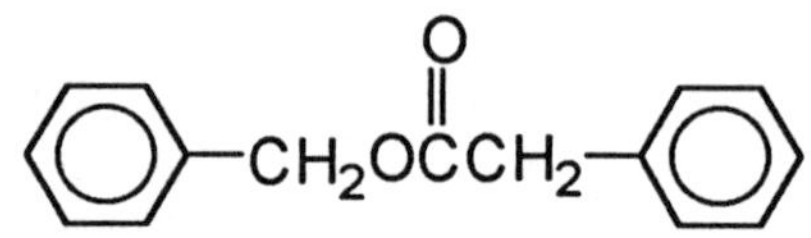

2. Which of the following compounds are named correctly?

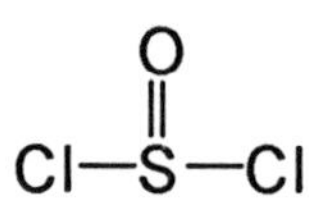

chlorosulfonic acid

I

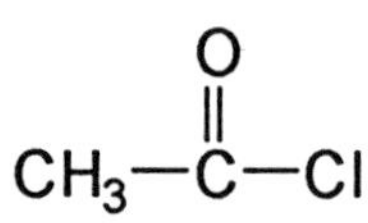

acetyl chloride

II

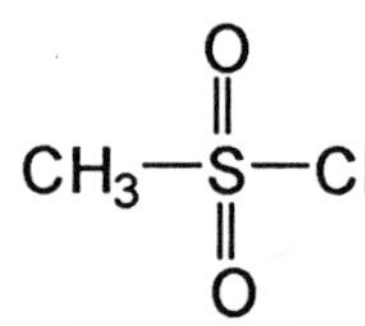

methanesulfonyl chloride

III

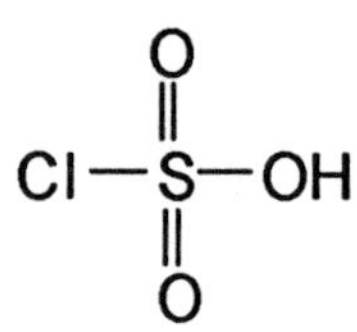

thionyl chloride

IV

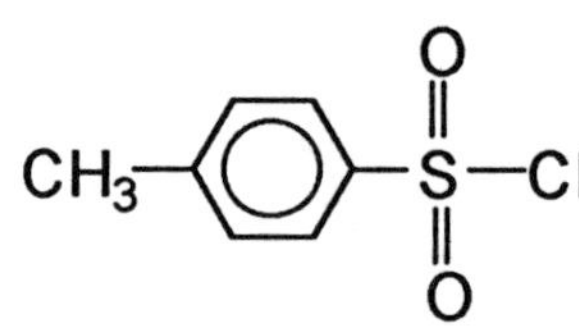

tosyl chloride

V

a) I, II, III   b) III, IV, V   c) I, III, V   d) II, III, V

3. Which of the following compounds are named correctly?

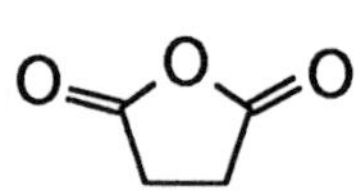

succinic anhydride

I

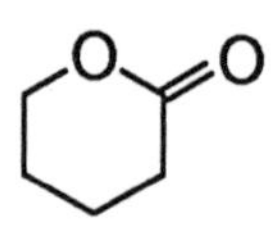

cyclohexyl anhydride

II

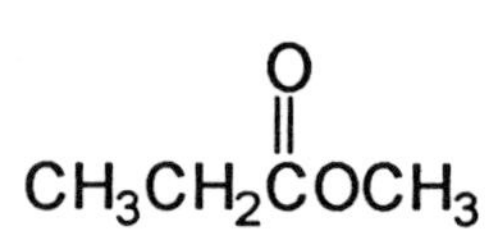

ethyl acetate

III

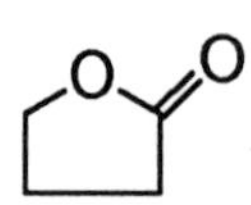

γ-butyrolactone

IV

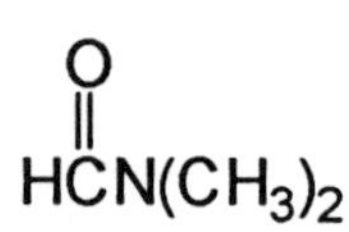

N,N-dimethyl formamide (DMF)

V

a) I, II, III   b) II, III, IV   c) I, IV, V   d) III, IV, V

4. What is the common name for the following structure?

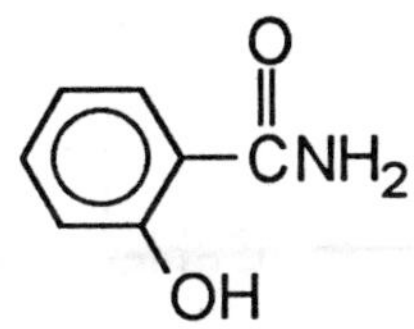

a) aspirin b) cyclosporin c) succinamide d) salicylamide

5. What is the order of decreasing reactivity for the indicated bonds in the following carboxylic acid derivatives (most reactive first)?

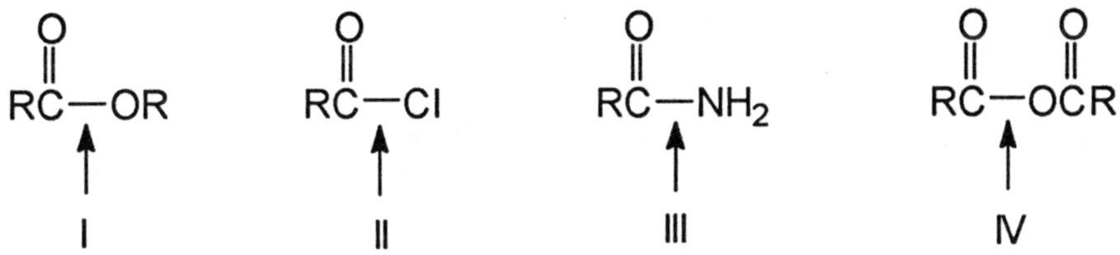

a) II, IV, I, III b) III, IV, I, II c) IV, I, II, III d) I, II, III, IV

6. Carboxylic acids and amides have in general higher boiling points than esters and anhydrides because of which property?

a) dipolar association
b) resonance stabilization
c) conjugated functional groups
d) hydrogen bonding

7. Which of the following reactions proceed as written?

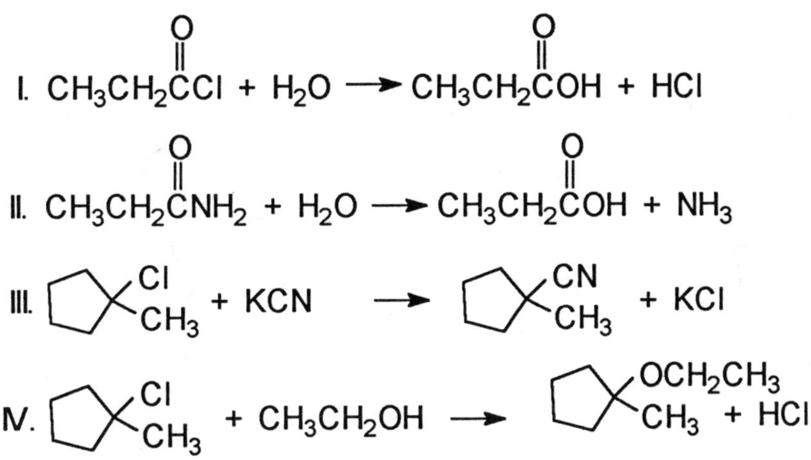

a) I, II b) I, IV c) II, III d) II, IV

8. The following reaction is fastest when Z is which group?

a) C_6H_5O- b) CH_3- c) CH_3O- d) $CH_3O\overset{O}{\overset{\|}{C}}O-$

9. What are the best conditions for the following preparation?

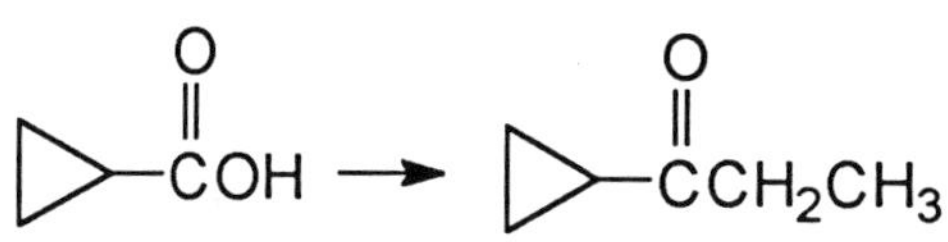

a) $SOCl_2$; $(CH_3CH_2)_2Cd$

b) $SOCl_2$; CH_3CH_2MgBr

c) $LiALH_4$: $SOCl_2$: Mg , ether ; CH_3CHO

d) CH_3OH , $H_3O^{\oplus}$; $LiAlH_4$

10. What is the principal product from the following reaction?

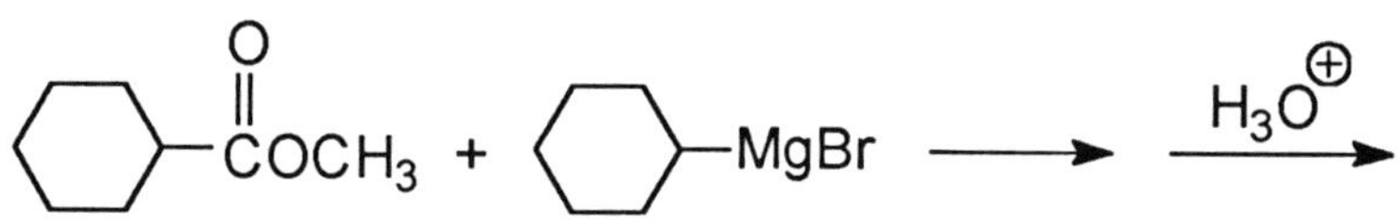

a) 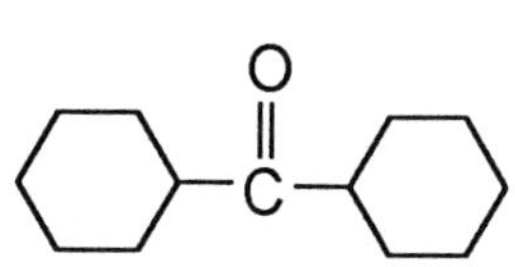

b) 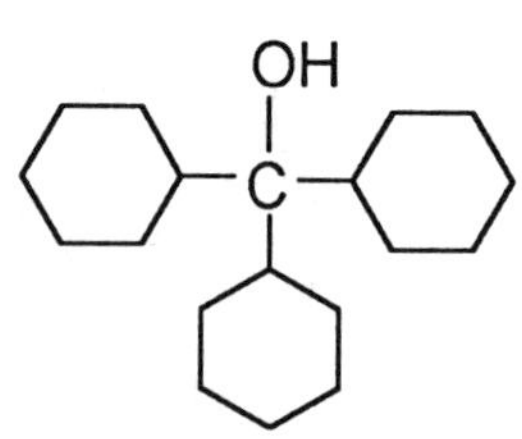

c) 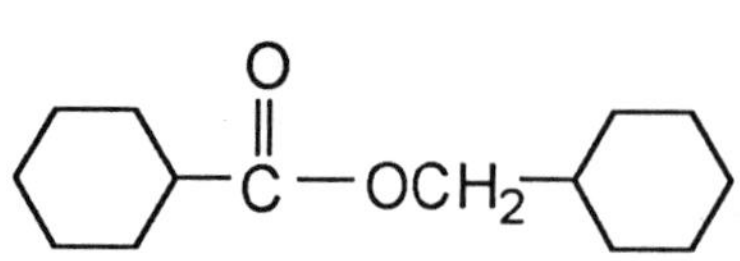

d) 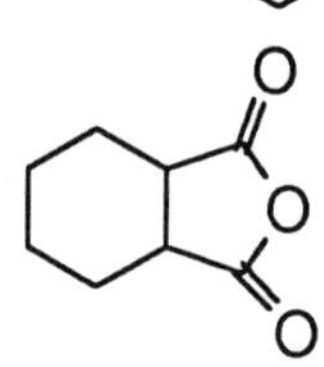

11. What are the best conditions for the following transformation?

$$CH_3\overset{\overset{O}{\|}}{C}CH_2CH_2\overset{\overset{O}{\|}}{C}OCH_3 \longrightarrow CH_3\overset{\overset{O}{\|}}{C}CH_2CH_2\overset{\overset{O}{\|}}{C}H$$

a) $\xrightarrow{LiAlH_4} \xrightarrow[HCl]{H_2O}$ b) $\xrightarrow{NaBH_4} \xrightarrow[HCl]{H_2O}$

c) $\xrightarrow[-78^{\circ}C]{DIBAL,\ toluene} \xrightarrow[HCl]{H_2O}$ d) $\xrightarrow[quinoline]{Pd,\ BaSO_4}$

12. What are the best conditions for the following reaction?

$$CH_3\overset{\overset{O}{\|}}{C}CH_2CH_2\overset{\overset{O}{\|}}{C}OCH_3 \longrightarrow CH_3\overset{\overset{OH}{|}}{C}CH_2CH_2\overset{\overset{O}{\|}}{C}OCH_3$$

a) $\xrightarrow{LiAlH_4} \xrightarrow[HCl]{H_2O}$ b) $\xrightarrow{NaBH_4} \xrightarrow[HCl]{H_2O}$

c) $\xrightarrow[-78^{\circ}C]{DIBAL,\ toluene} \xrightarrow[HCl]{H_2O}$ d) $\xrightarrow[quinoline]{Pd,\ BaSO_4}$

13. What are the best conditions for the following reaction?

$$CH_3\overset{\overset{O}{\|}}{C}CH_2CH_2\overset{\overset{O}{\|}}{C}OCH_3 \longrightarrow CH_3\overset{\overset{OH}{|}}{C}CH_2CH_2CH_2OH$$

a) $\xrightarrow{LiAlH_4} \xrightarrow[HCl]{H_2O}$ b) $\xrightarrow{NaBH_4} \xrightarrow[HCl]{H_2O}$

c) $\xrightarrow[-78^{\circ}C]{DIBAL,\ toluene} \xrightarrow[HCl]{H_2O}$ d) $\xrightarrow[quinoline]{Pd,\ BaSO_4}$

14. What is the principal product from the following reaction?

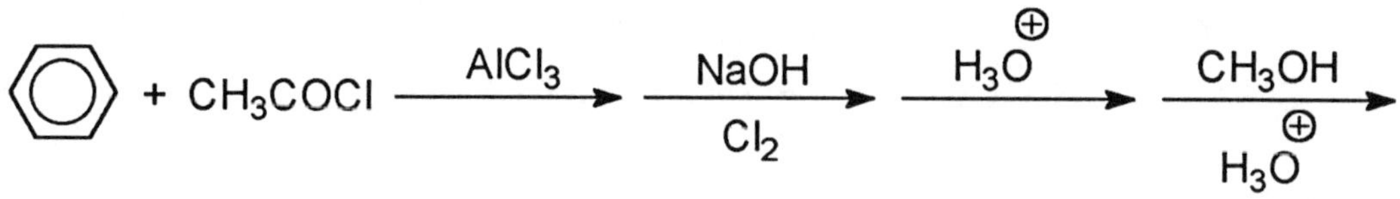

a) 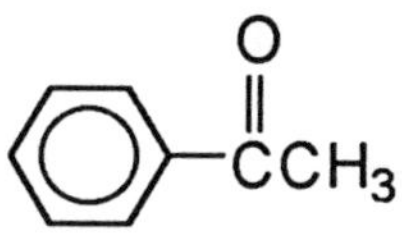

b) 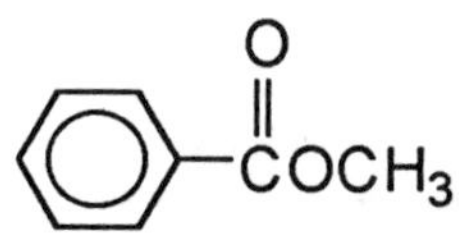

c) 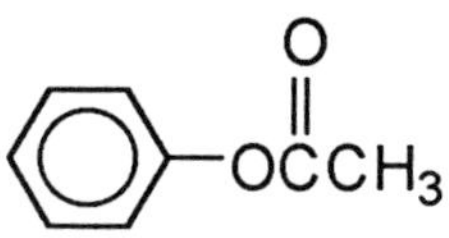

d) 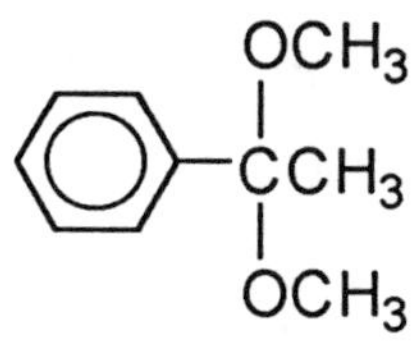

15. What are the best conditions for the following synthesis?

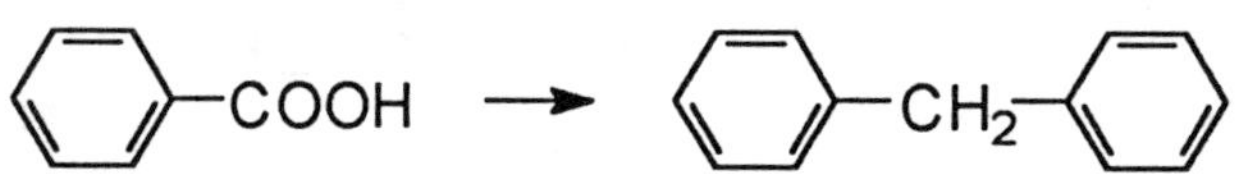

a) $\xrightarrow{SOCl_2}$ $\xrightarrow[C_6H_{12}]{AlCl_3}$ $\xrightarrow[HCl]{Zn(Hg)}$

b) $\xrightarrow{SOCl_2}$ $\xrightarrow[ether]{C_6H_{11}MgBr}$

c) $\xrightarrow{SOCl_2}$ $\xrightarrow[C_6H_{12}]{AlCl_3}$ $\xrightarrow{(C_6H_{11})_2Cd}$

d) $\xrightarrow{LiAlH_4}$ $\xrightarrow{SOCl_2}$ $\xrightarrow[ether]{Mg}$ $\xrightarrow{H_2O}$ $\xrightarrow[AlCl_3]{C_6H_{12}}$

16. What is the principal product when ethylene diamine is heated with dimethyloxalate?

a) $CH_3\overset{O}{\overset{\|}{C}}NHCH_2CH_2NH\overset{O}{\overset{\|}{C}}CH_3$

b) 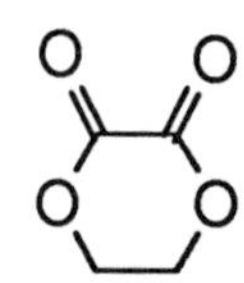

c) 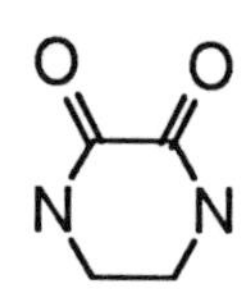

d) $CH_3\overset{O}{\overset{\|}{C}}NHCH(CH_3)NH\overset{O}{\overset{\|}{C}}CH_3$

17. What is the principal product from the following reaction?

$$CH_3CH_2O\overset{O}{\overset{\|}{C}}(CH_2)_5\overset{O}{\overset{\|}{C}}OCH_2CH_3 \xrightarrow{NaOCH_2CH_3} \xrightarrow{H_3\overset{\oplus}{O}}$$

a) 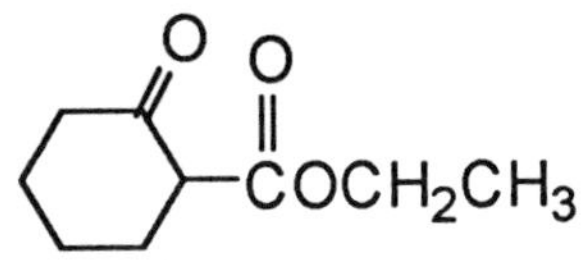

b) 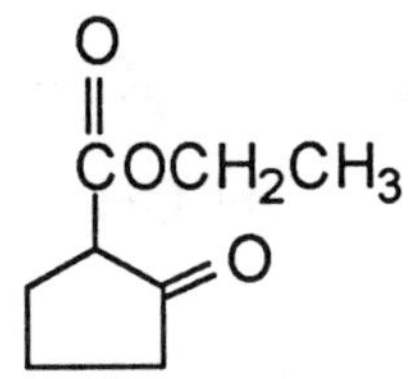

c) 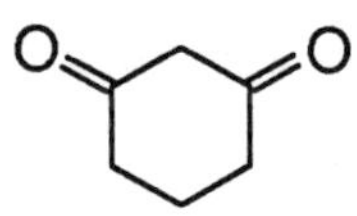

d) 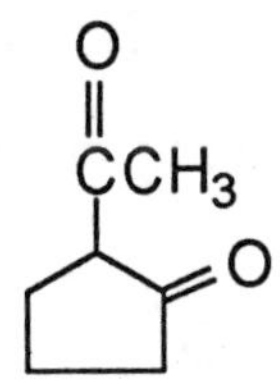

18. What is the principal product from the following reaction?

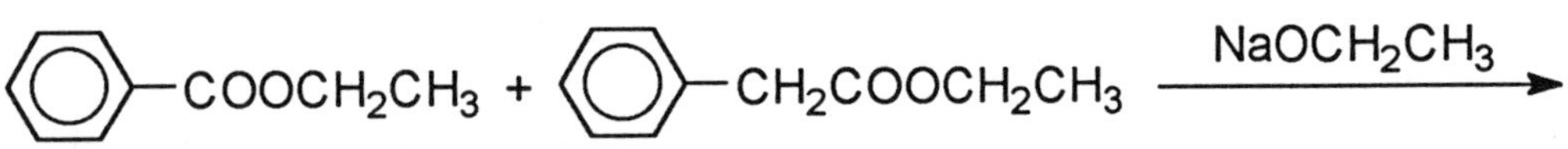

a) 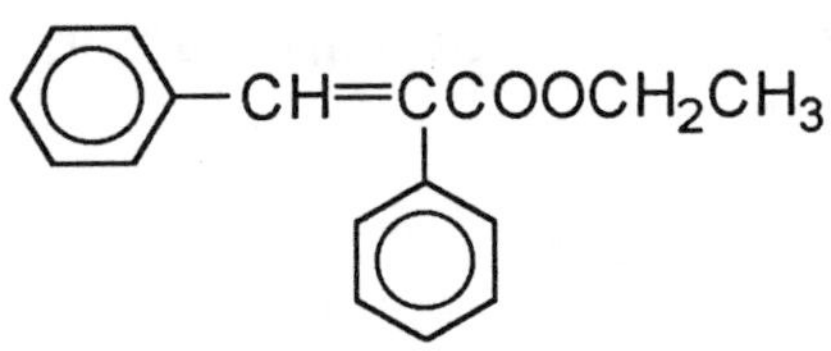

b) 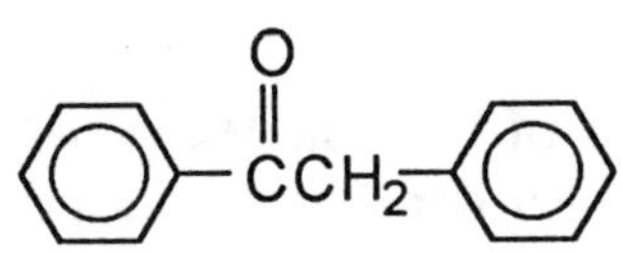

c) 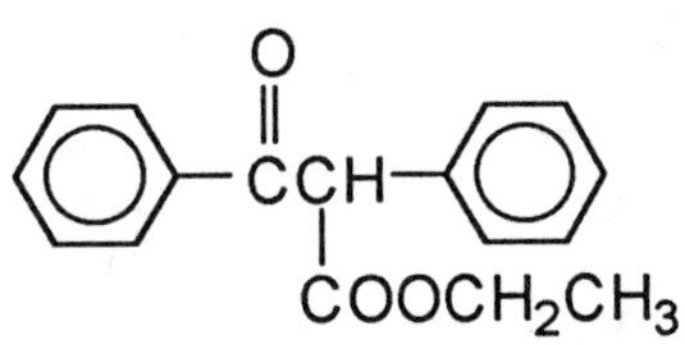

d) 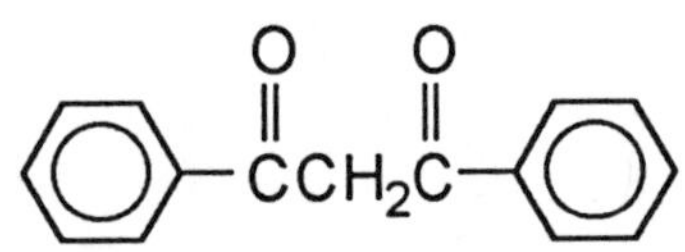

19. What are the best conditions for the following reaction?

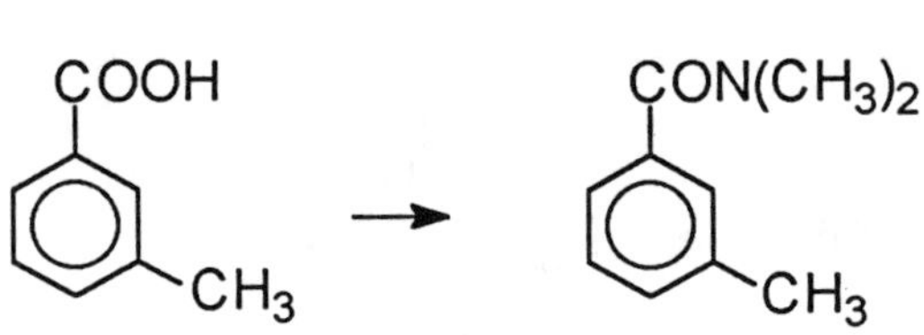

a) $\xrightarrow{SOCl_2} \xrightarrow{(CH_3)_2NH}$

b) $\xrightarrow{NaOH} \xrightarrow{(CH_3)_2NH}$

c) $\xrightarrow{NH_3} \xrightarrow{CH_3CH_2Br}$

d) $\xrightarrow[\text{room temperature}]{(CH_3)_2NH}$

20. What is the principal product from the following reaction?

$$CH_3CH_2CN + CH_3CH_2MgBr \xrightarrow{\text{ether}} \xrightarrow[H_2O]{HCl}$$

a) $CH_3CH_2\overset{OH}{\underset{CH_2CH_3}{C}}CH_2CH_3$ b) $CH_3CH_2\overset{NH_2}{CH}CH_2CH_3$

c) $CH_3CH_2\overset{O}{\overset{\|}{C}}NHCH_2CH_3$ d) $CH_3CH_2\overset{C}{\overset{\|}{C}}CH_2CH_3$

21. Which of the following compounds will yield benzoic acid when hydrolyzed?

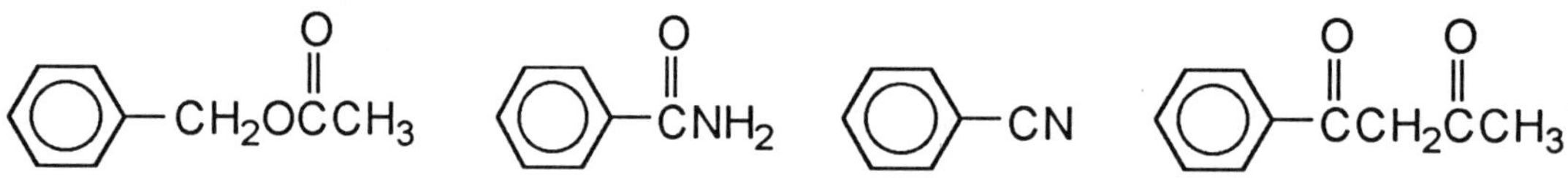

I II III IV

a) I, II b) III, IV c) II, III d) II, IV

22. Which of the following represents the formation of an important intermediate in the esterification of acetic acid by methanol?

a) $CH_3CH_2COOH + CH_3OH \longrightarrow CH_3CH_2\overset{O^{\ominus}}{\underset{CH_3\overset{\oplus}{O}H}{C}}OH$

b) $CH_3CH_2COOH + CH_3O^{\ominus} \longrightarrow CH_3CH_2\overset{O^{\ominus}}{\underset{CH_3O}{C}}OH$

c) $CH_3CH_2\overset{\overset{\oplus}{O}H}{\overset{\|}{C}}OH + CH_3OH \longrightarrow CH_3CH_2\overset{OH}{\underset{CH_3\overset{\oplus}{O}H}{C}}OH$

d) $CH_3CH_2\overset{O}{\overset{\|}{C}}OH + CH_3OH \longrightarrow CH_3CH_2\overset{O^{\ominus}}{\underset{CH_3\overset{\oplus}{O}H}{C}}OH$

23. What is the product from the following reaction?

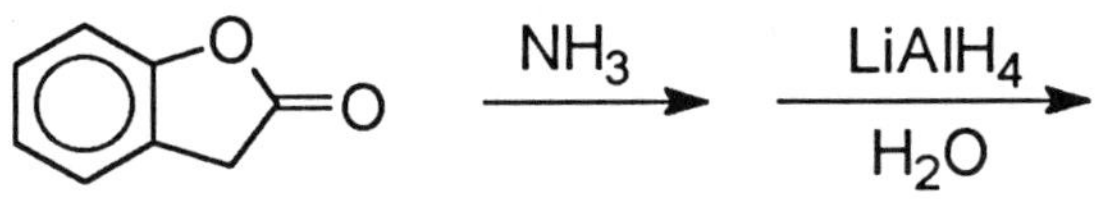

a) 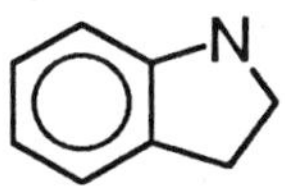 b) 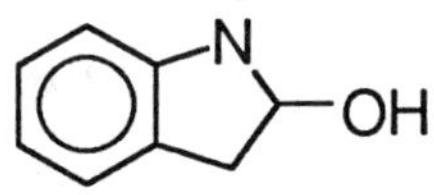 c) 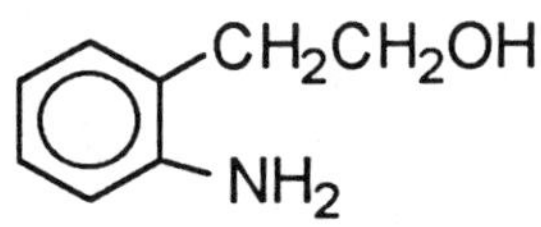 d) 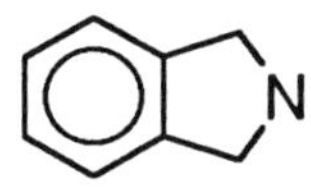

24. Which of the following compounds does not yield nicotinamide (niacin) when reacted with ammonia?

a) 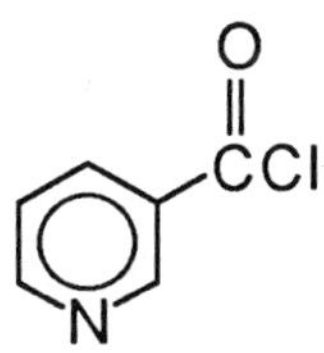 b) 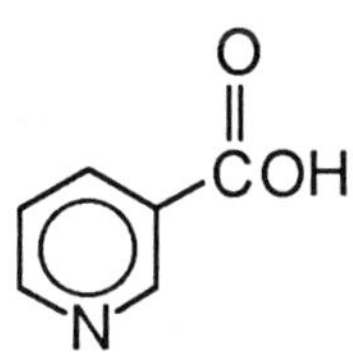 c) 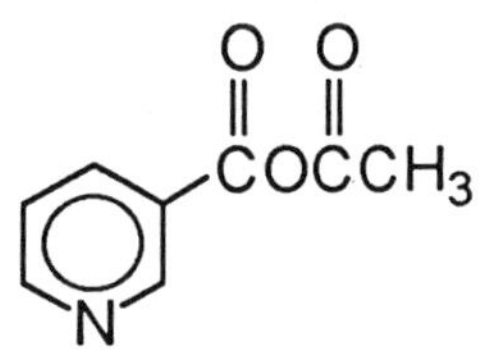 d) 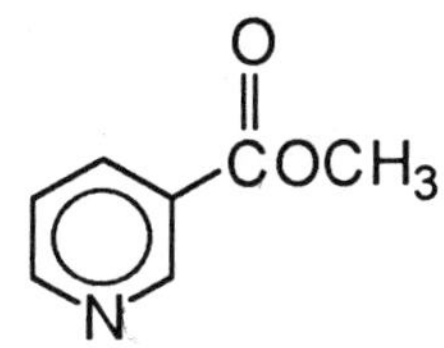

25. What is the principal product from the following reaction?

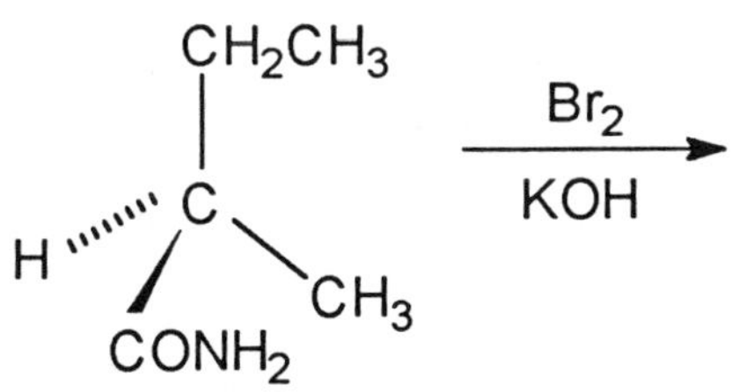

a) 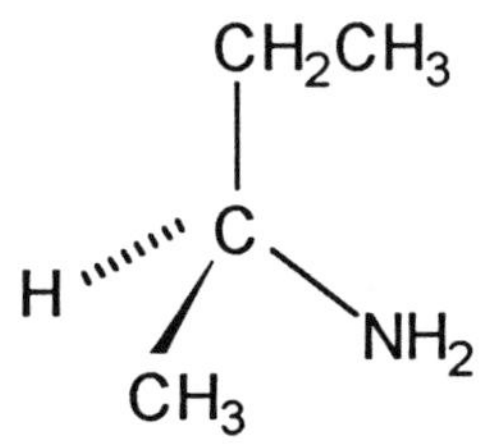 b) 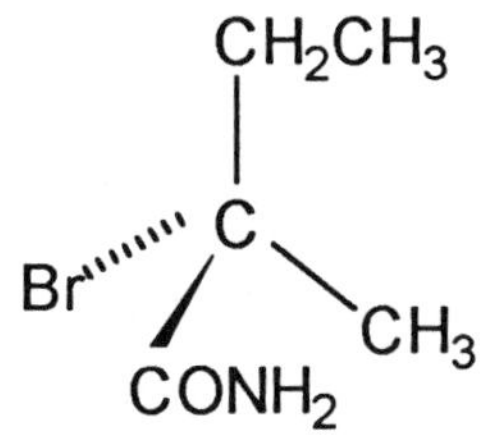

c) 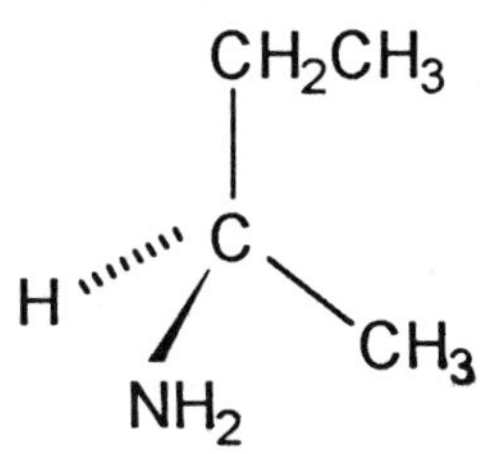 d) 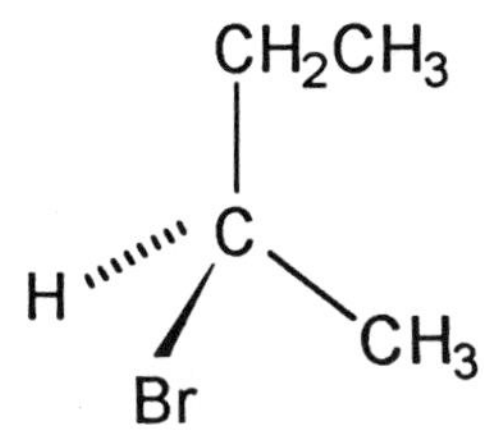

Answers

1. a)
2. d)
3. c)
4. d)
5. a)
6. d)
7. b)
8. d)
9. a)
10. b)
11. c)
12. b)
13. a)
14. b)
15. a)
16. c)
17. a)
18. c)
19. a)
20. d)
21. c)
22. c)
23. d)
24. b)
25. c)

Chapter 21 Enolate Anions and Enamines

1. What are the major products when the following ketones are converted to the enolate ions under kinetic control and then reacted with trimethylsilyl chloride?

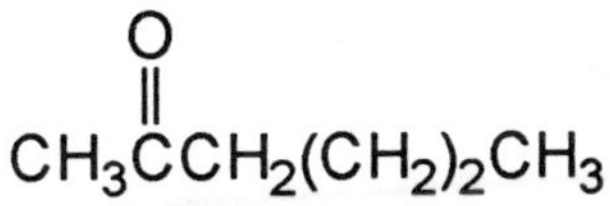

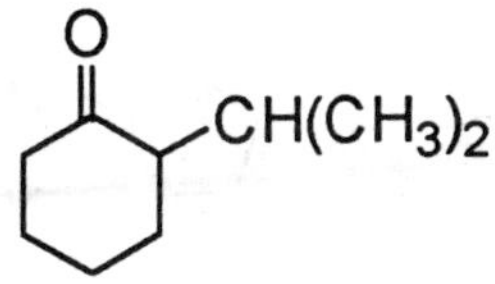

I. 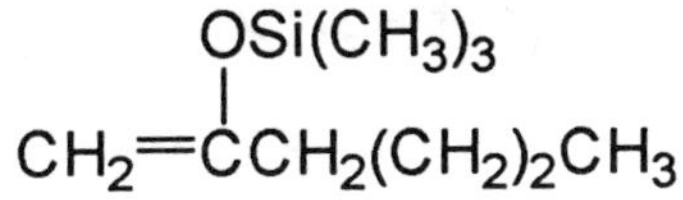

II. 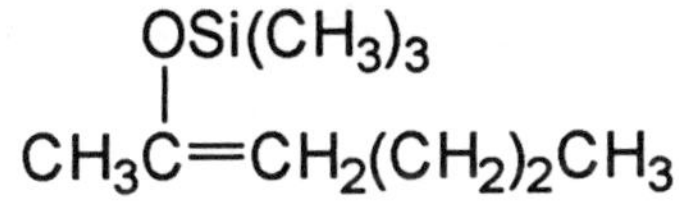

III. 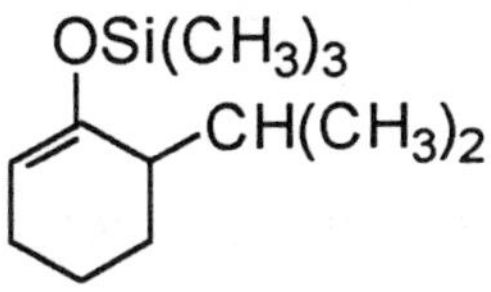

IV. 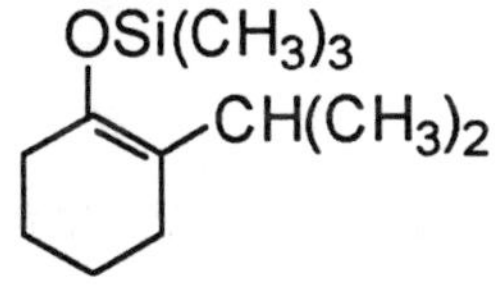

a) I, III b) II, III c) I, IV d) II, IV

2. What are the major products when the following ketones are converted to the enolate ions under thermodynamic control and then reacted with trimethylsilyl chloride?

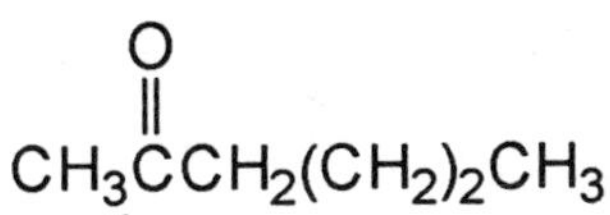

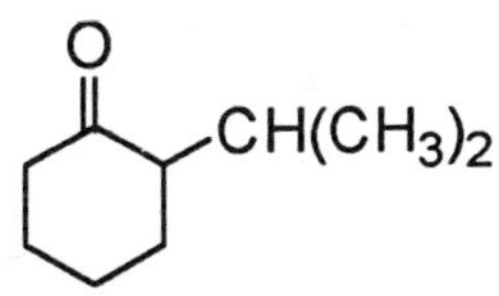

I. 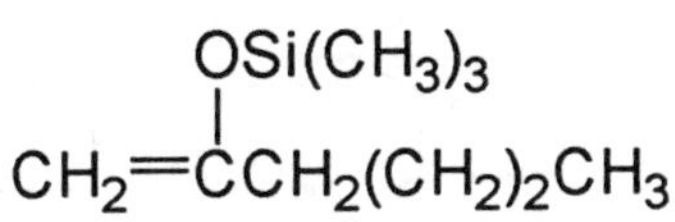

II. 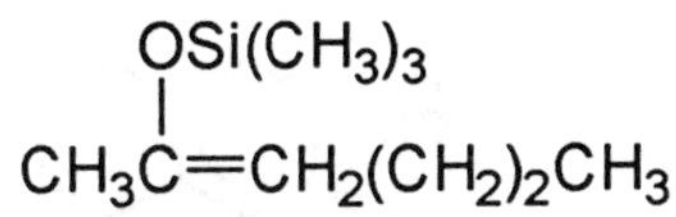

III. 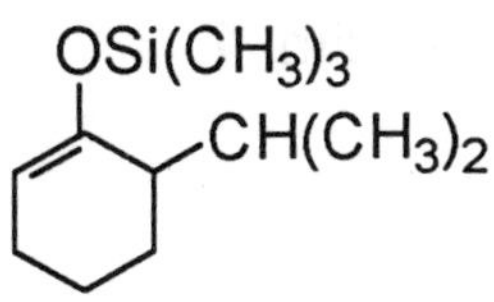

IV. 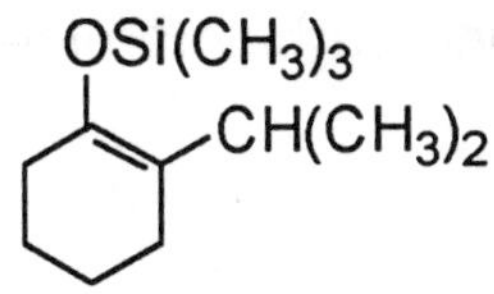

a) I, III b) II, III c) I, IV d) II, IV

Chapter 21 Enolate Anions and Enamines

3. What are the major reactants for the directed aldol condensation of the following product?

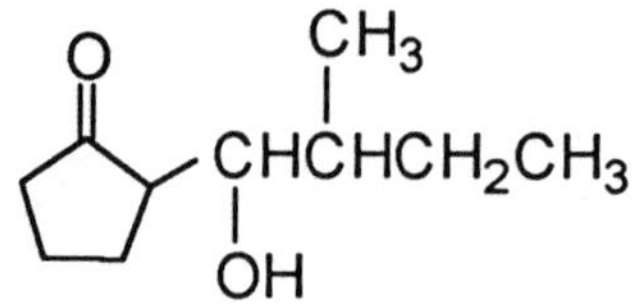

a) $CH_3CH_2\overset{CH_3}{\overset{|}{C}}HCHO$ + 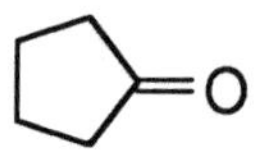

b) $CH_3\overset{CH_3}{\overset{|}{C}}HCH_2CHO$ + 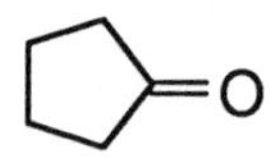

c) $CH_3CH_2\overset{O}{\overset{||}{C}}CH_3$ + 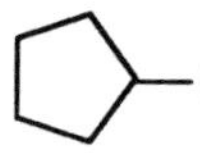CHO

d) $CH_3CH_2CH_2CHO$ + 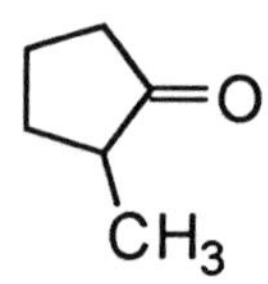

4. What are the reactants in the enamine synthesis of the following compound?

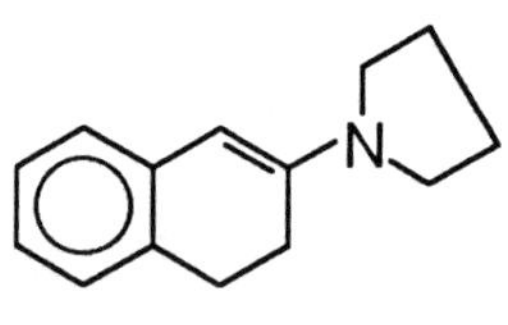

a) 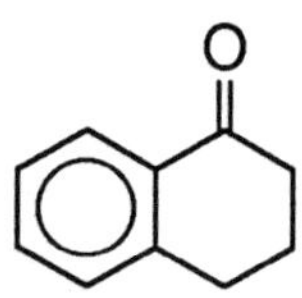 + 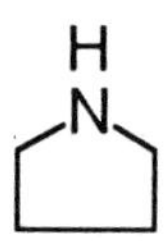

b) 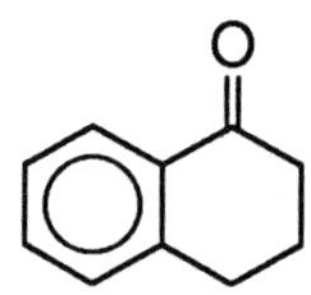 + 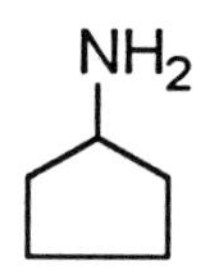

c) 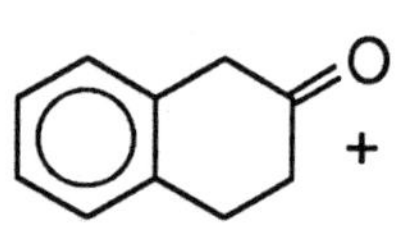 + 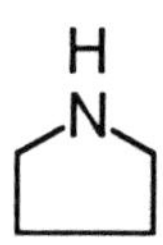

d) 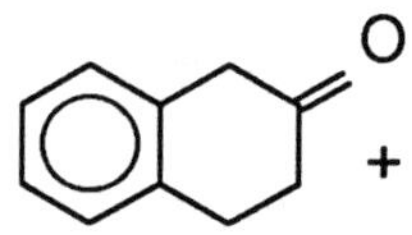 + 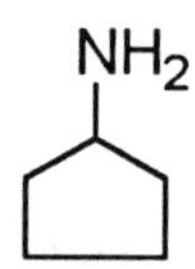

5. Which carbinolamine is formed in the following reaction?

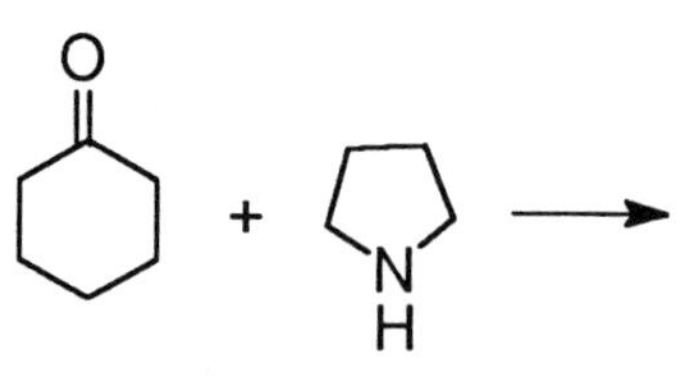

a) 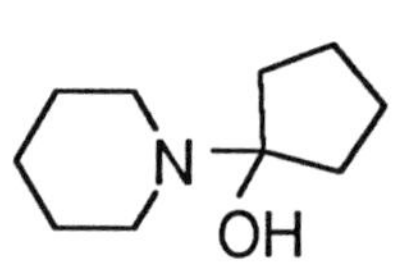

b) 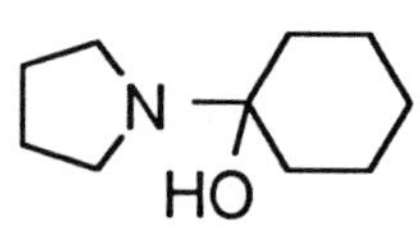

c) 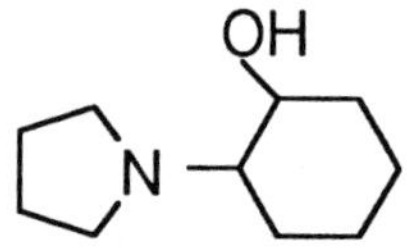

d) 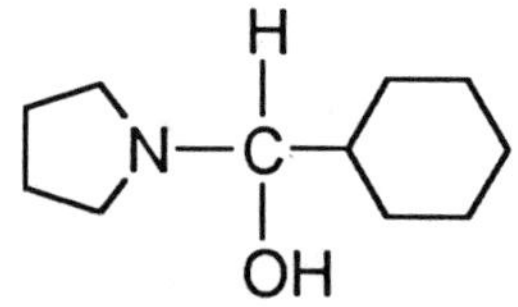

6. Indicate which of the products will not be formed in a synthesis starting with the following enamine and with chloroacetone or allyl chloride or methyl chloroacetate.

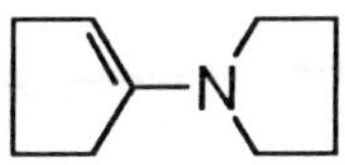

a) 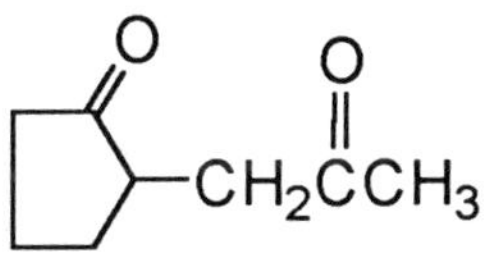

b) 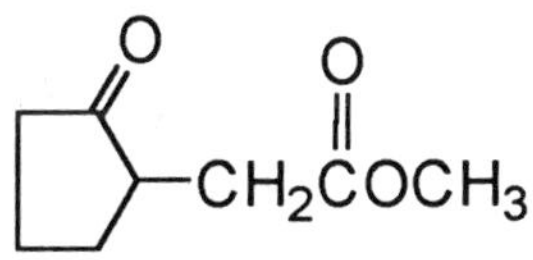

c) 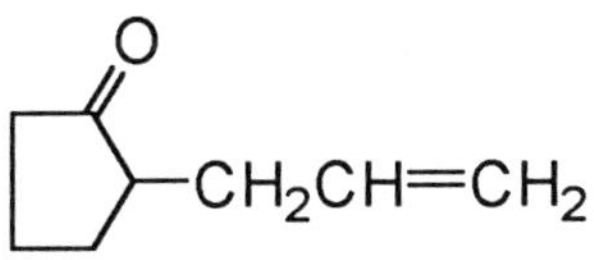

d) 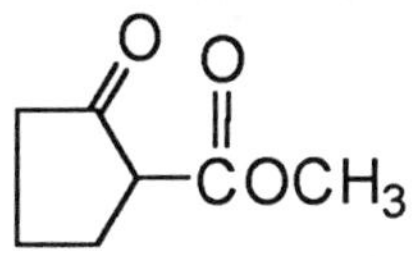

7. What is the major product from the following reactions?

CH_3CH_2CHO + 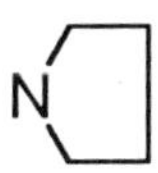 ⟶ $\xrightarrow{CH_2{=}CHCH_2Cl}$ $\xrightarrow[HCl]{H_2O}$

a) $CH_2{=}CHCH_2CH(CH_3)CHO$

b) $CH_3CH_2CH{=}C(CH_3)CHO$

c) $CH_2{=}CHCH(CH_3)CH_2CHO$

d) $CH_2{=}CHCH_2CH_2CH_2CHO$

8. Which of the following halides can not be used effectively in an acetoacetic ester synthesis?

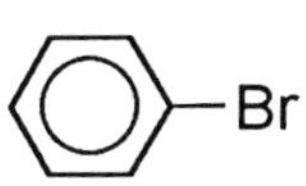	$CH_2{=}CHCH_2Br$	$(CH_3)_3CBr$	CH_3I
I	II	III	IV

a) I, II b) III, IV c) II, IV d) I, III

9. What is the order of decreasing acidity for the indicated protons (most acidic first)?

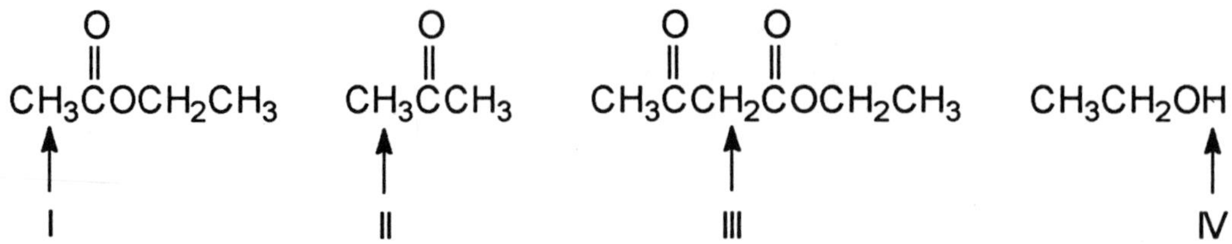

a) I, II, III, IV   b) III, IV, II, I   c) II, I, III, IV   d) IV, III, II, I

10. Which halide can be used in an acetoacetic ester synthesis of 2-hexanone?

a) $CH_3CH_2CH_2Br$ b) $CH_3C(=O)CH_2Cl$

c) $CH_2{=}CHCH_2Cl$ d) CH_3CH_2Br + CH_3I

11. What is the major product from the following series of reactions?

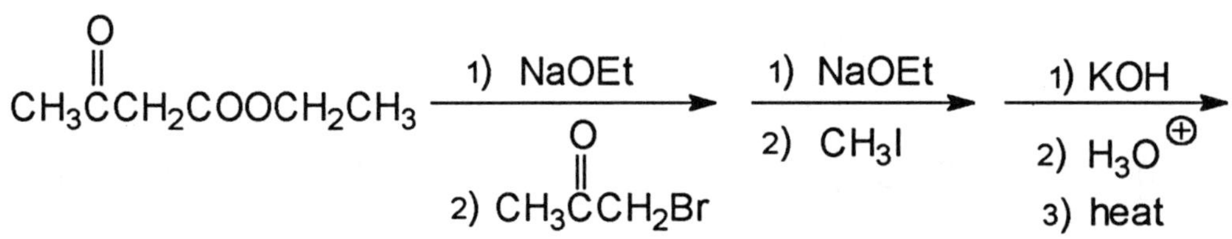

a) $CH_3C(=O)CH(CH_3)CH_2C(=O)CH_3$ b) $CH_3CH(CH_3)C(=O)CH_2C(=O)CH_3$

c) $CH_3C(=O)CH(CH_3)CH(CH_3)C(=O)CH_3$ d) $CH_3C(=O)CH_2CH_2C(=O)CH_3$

12. What is the major product from the following series of reactions?

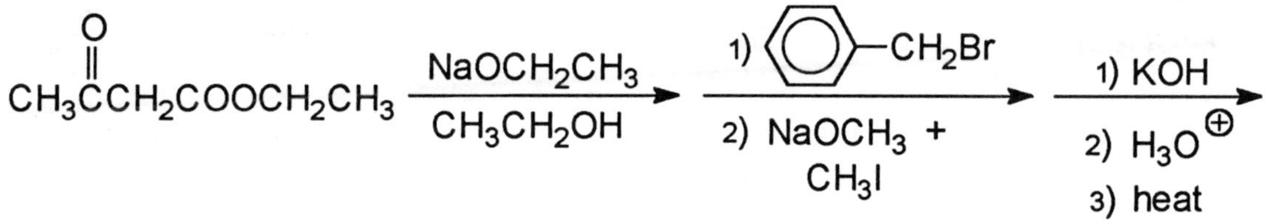

a) 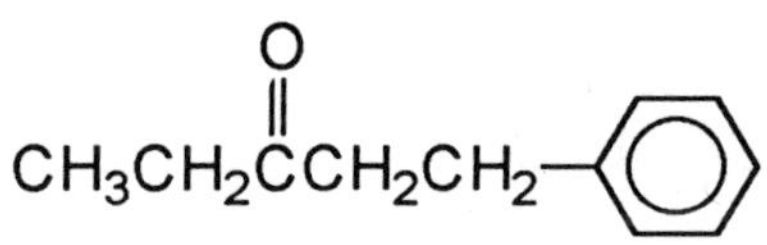

b) 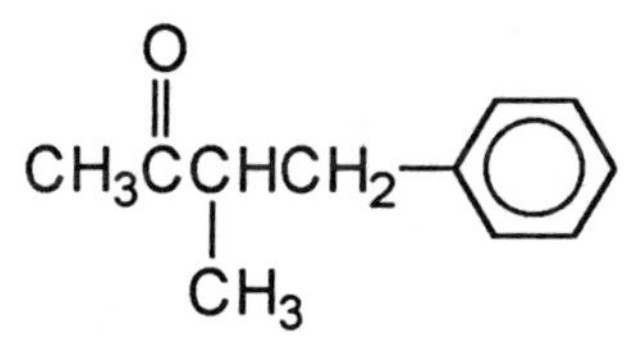

c) 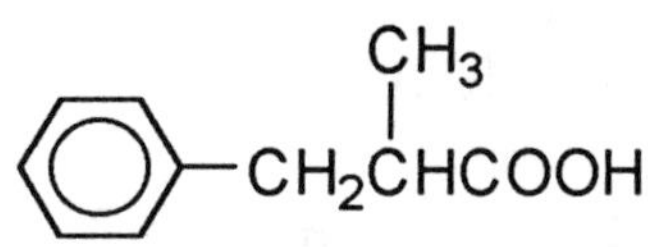

d) 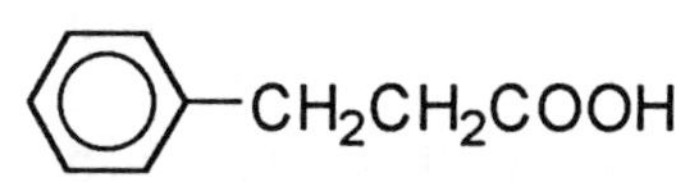

13. The malonic ester synthesis works with which of the following halides?

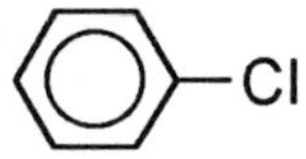

I

$CH_2{=}CHCH_2Br$

II

$(CH_3)_3CBr$

III

CH_3I

IV

a) I, II b) III, IV c) II, IV d) I, III

14. Which halide can be used with diethylmalonate to prepare hexanoic acid?

a) CH_3CH_2Br

b) $CH_3CH_2CH_2Br$

c) $CH_3CH_2CH_2CH_2Br$

d) $CH_3CH_2CH_2CH_2CH_2Br$

15. What is the major product from the following series of reactions?

$$BrCH_2CH_2Br + 2\ CH_3CH_2OOC\overset{\ominus}{C}HCOOCH_2CH_3\ Na^{\oplus} \longrightarrow \xrightarrow[\text{2) } H_3O^{\oplus}\text{ 3) heat}]{\text{1) KOH}}$$

a) 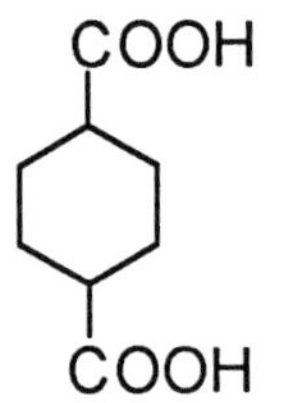

b) 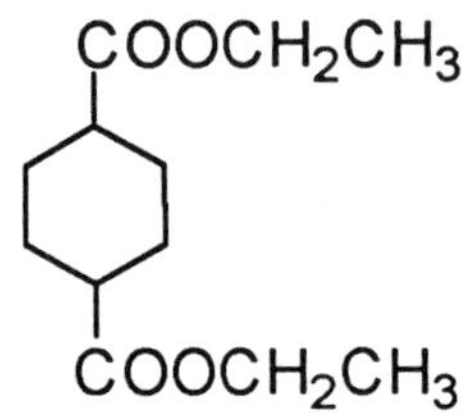

c) $HOOC(CH_2)_4COOH$

d) $CH_3\overset{O}{\overset{\|}{C}}CH_2CH_2\overset{O}{\overset{\|}{C}}CH_3$

16. What is the major product from the following series of reactions?

$$CH_3CH_2O\overset{O}{\overset{\|}{C}}CH_2\overset{O}{\overset{\|}{C}}OCH_2CH_3 \xrightarrow[\text{2) } CH_3CH_2I]{\text{1) } NaOCH_2CH_3} \xrightarrow[\text{2) } CH_3I]{\text{1) } NaOCH_2CH_3} \xrightarrow[\text{2) } H_3O^{\oplus}\text{ 3) heat}]{\text{1) KOH}}$$

a) $CH_3CH_2\underset{}{C}H(CH_3)COOH$

b) $CH_3CH(CH_3)CH_2COOH$

c) $CH_3CH_2CH(CH_2CH_3)COOH$

d) $CH_3CH_2CH_2CH_2COOH$

17. Which of the following molecules can become effective nucleophiles for the Michael reaction?

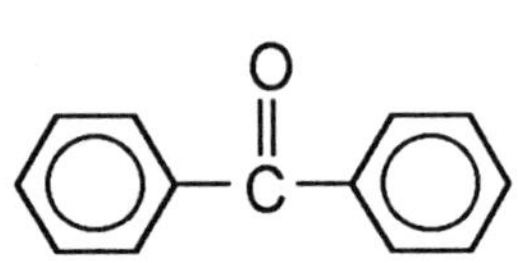

I

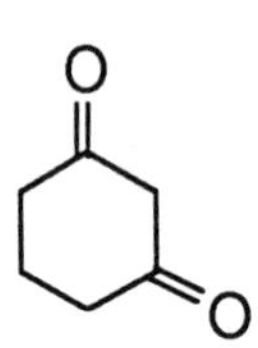

II

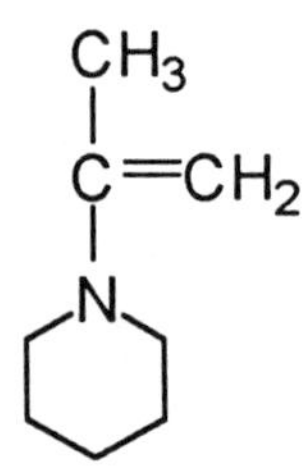

III

$CH_2{=}CH\overset{O}{\overset{\|}{C}}COOCH_2CH_3$

IV

a) I, II b) II, III c) III, IV d) I, IV

18. Which of the following molecules can act as nucleophile acceptors for the Michael reaction?

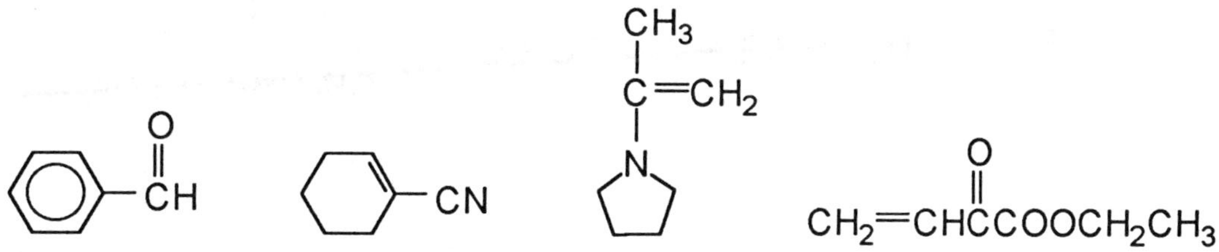

I II III IV

a) I, II b) III, IV c) II, IV d) I, III

19. What is the major product from the following reaction?

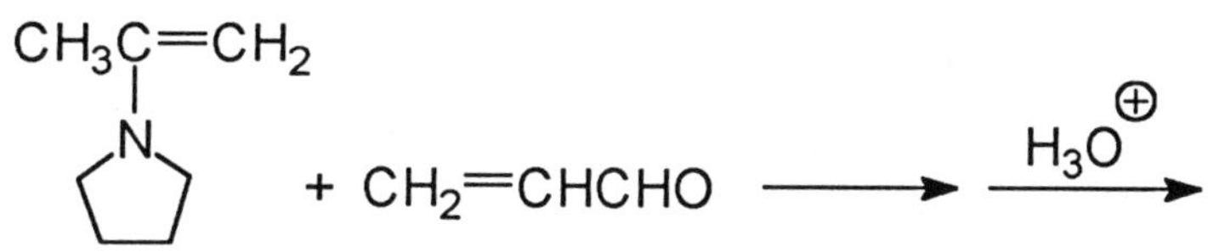

a) $HC(=O)CH_2CH_2CH_2CH(=O)$

b) $CH_3C(=O)CH_2CH_2CH_2CH(=O)$

c) $CH_3CH(CH_3)C(=O)CH_2CH(=O)$

d) 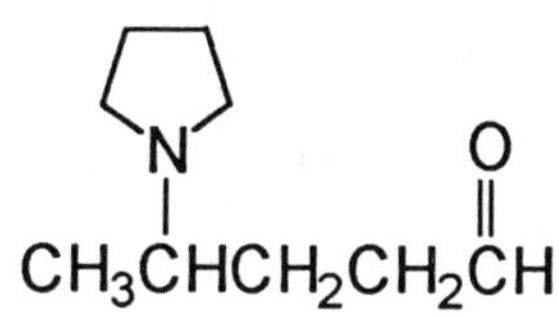

20. What is the major product from the following acetoacetic ester synthesis?

$$CH_3C(=O)CH_2CO_2Et + CH_2{=}CHC(=O)OEt \xrightarrow[EtOH]{NaOEt} \xrightarrow[\text{2) } H_3O^{\oplus} \text{ 3) heat}]{\text{1) KOH}}$$

a) $CH_3CO(CH_2)_3COOH$ b) $CH_3CO(CH_2)_4COOH$

c) $CH_3CO(CH_2)_2COOH$ d) $HOOC(CH_2)_3COOH$

21. What is the major product from the following reaction?

$$CH_2(\overset{O}{\overset{\|}{C}}OCH_2CH_3)_2 + CH_2{=}CH\overset{O}{\overset{\|}{C}}OCH_2CH_3 \xrightarrow[CH_3CH_2OH]{NaOCH_2CH_3} \xrightarrow[\substack{2)\ H_3O^{\oplus} \\ 3)\ heat}]{1)\ KOH}$$

a) $CH_3CO(CH_2)_3COOH$ b) $CH_3CO(CH_2)_4COOH$

c) $CH_3CO(CH_2)_2COOH$ d) $HOOC(CH_2)_3COOH$

22. What reactants can be used to prepare the following barbiturate?

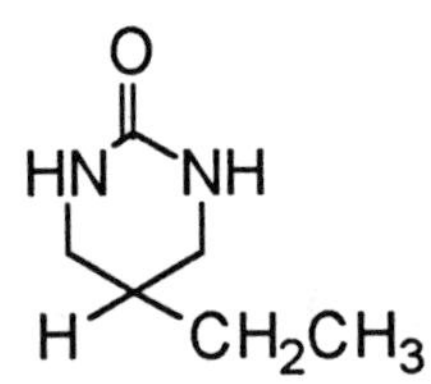

I. $CH_3\overset{O}{\overset{\|}{C}}CH_2\overset{O}{\overset{\|}{C}}OCH_2CH_3$ II. CH_3CH_2Br

III. $CH_3CH_2O\overset{O}{\overset{\|}{C}}CH_2\overset{O}{\overset{\|}{C}}OCH_2CH_3$ IV. $H_2N\overset{O}{\overset{\|}{C}}NH_2$

a) I, III, IV b) I, II, III c) I, II, IV d) II, III, IV

23. Ethyl benzoate is reacted with ethyl acetate in a solution of sodium ethoxide. The product is treated with sodium ethoxide and ethyl bromide, followed by reaction with aqueous acid and then heat. What is the most reasonable product from these reactions?

a) 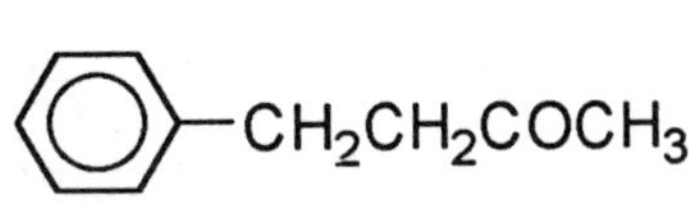 b) 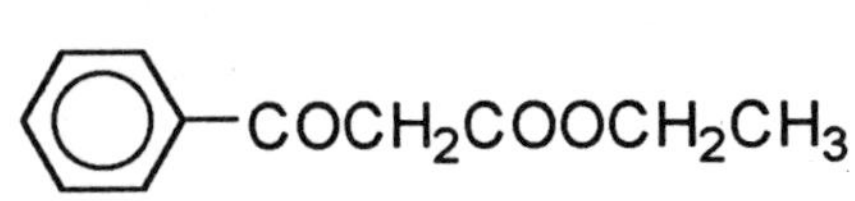

c) 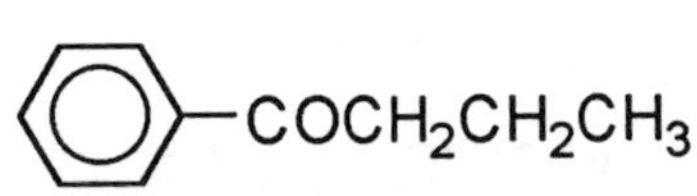 d) 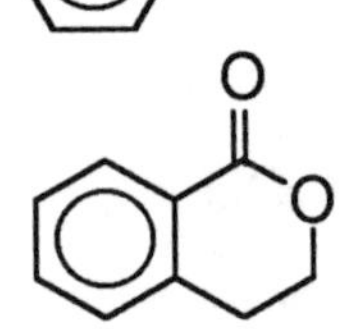

24. What is the best reaction route for the following synthesis?

$$CH_2(COOEt)_2 \longrightarrow HO(CH_2)_5OH$$

a) $\xrightarrow{\text{NaOEt}} \xrightarrow{CH_2{=}C(CH_3)COOEt} \xrightarrow[\text{3) heat}]{\text{1) KOH; 2) } H_3O^{\oplus}} \xrightarrow{LiAlH_4}$

b) $\xrightarrow{\text{NaOEt}} \xrightarrow{CH_2{=}CHCOCH_3} \xrightarrow[\text{3) heat}]{\text{1) KOH; 2) } H_3O^{\oplus}} \xrightarrow{LiAlH_4}$

c) $\xrightarrow{\text{NaOEt}} \xrightarrow{Br(CH_2)_3OCH_2CH_3} \xrightarrow[\text{3) heat}]{\text{1) KOH; 2) } H_3O^{\oplus}} \xrightarrow{LiAlH_4}$

d) $\xrightarrow{\text{NaOEt}} \xrightarrow{CH_2{=}CHCN} \xrightarrow[\text{3) heat}]{\text{1) KOH; 2) } H_3O^{\oplus}} \xrightarrow{LiAlH_4}$

25. In order to prepare 4-methyl-5-phenyl-2-pentanone, which of the following halides is used to alkylate ethyl acetoacetate?

a) 2-bromo-1-phenylpropane
b) 1-bromo-2-methyl-3-phenylpropane
c) benzyl bromide and methyl bromide
d) 1-bromo-2-phenylethane and methyl bromide

Answers

1. a)
2. d)
3. a)
4. c)
5. b)
6. d)
7. a)
8. d)
9. b)
10. a)
11. a)
12. b)
13. c)
14. c)
15. c)
16. a)
17. b)
18. c)
19. b)
20. a)
21. d)
22. d)
23. c)
24. d)
25. a)

Chapter 22 Amines

1. What is the correct structure for 3-(N,N-dimethylamino)hexane?

a) 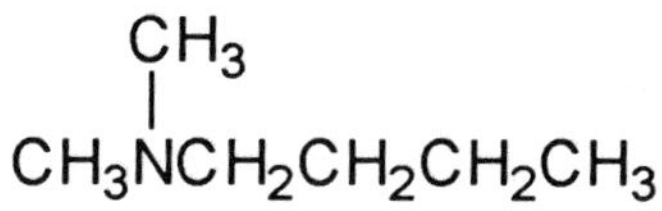

b) 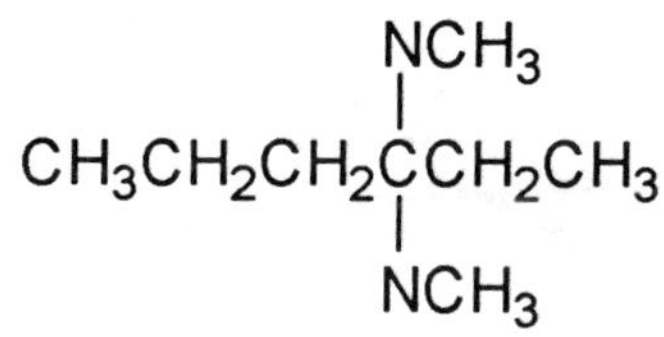

c) 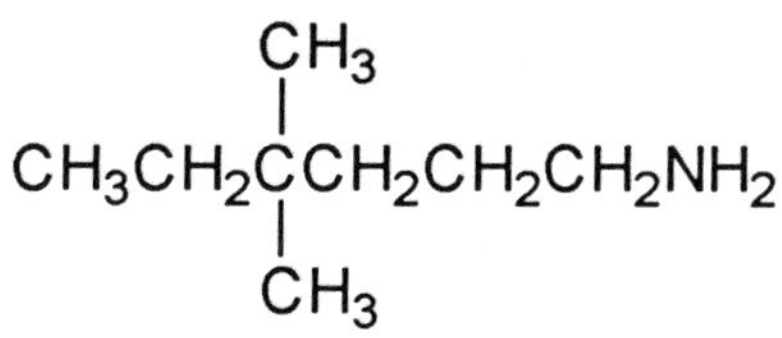

d) 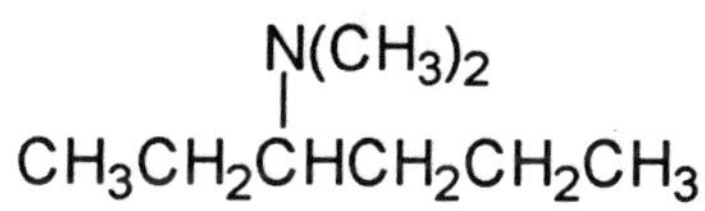

2. Which of the following matches of names and structures are correct?

I. piperidine — A. 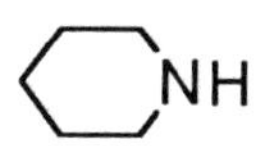

II. pyrrolidine — B. 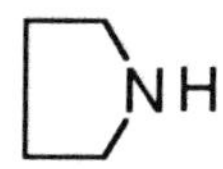

III. pyrrole — C. 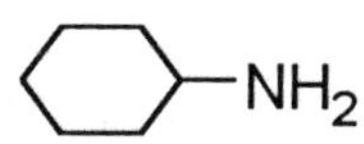

IV. pyridine — D. 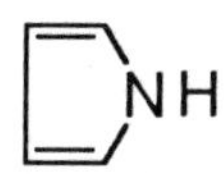

V. cyclohexylamine — E. 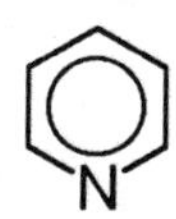

a) I and D, II and B
b) III and A, IV and E
c) V and C, III and B
d) III and D, IV and E

3. Which of the following structures are secondary amines?

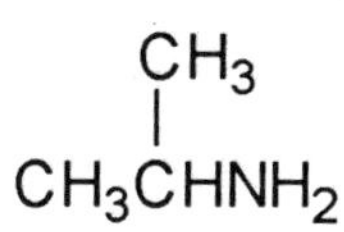
I

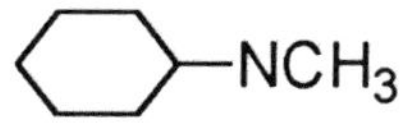
II

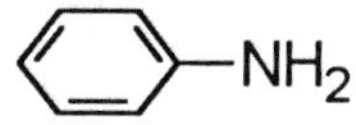
III

IV

a) I, II
b) II, III
c) III, IV
d) II, IV

4. Which of the following reactions will proceed as written?

I. $CH_3NH_2 + H_2O \longrightarrow CH_3NH_3^{\oplus} + OH^{\ominus}$

II. $CH_3NH_2 + CH_3COOH \longrightarrow CH_3NH_3^{\oplus} + CH_3COO^{\ominus}$

III. $CH_3NH_2 + (CH_3)_3N \longrightarrow CH_3NH^{\ominus} + (CH_3)_3NH^{\oplus}$

IV. 

a) I, II b) III, IV c) II, IV d) I, III

5. When the following compounds are dissolved in water at the same concentrations, which one will have the lowest pH?

a) 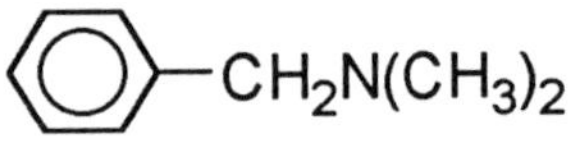

b) 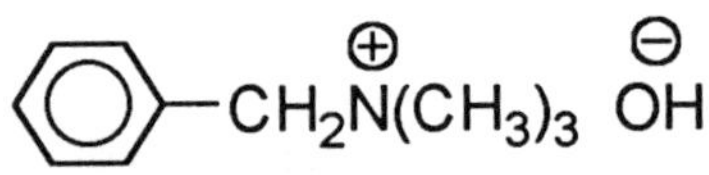

c) 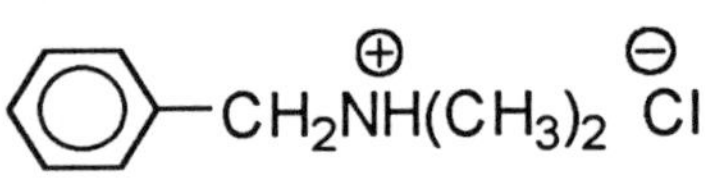

d) 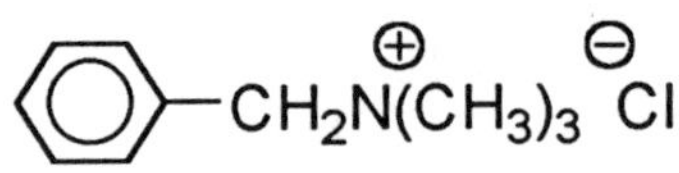

6. What is the order of increasing basicity for the following amines (weakest first)?

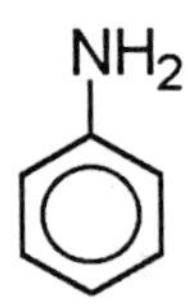

I

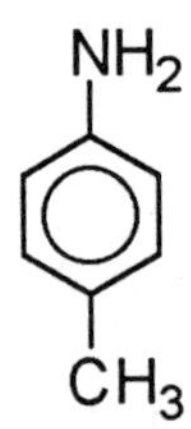

II

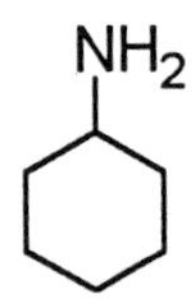

III

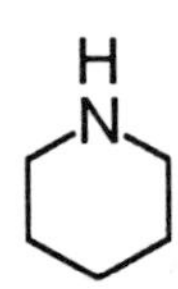

IV

a) IV, II, III, I b) II, I, III, IV c) I, II, III, IV d) II, I, IV, III

7. Which of the following compounds is the strongest base?

a) 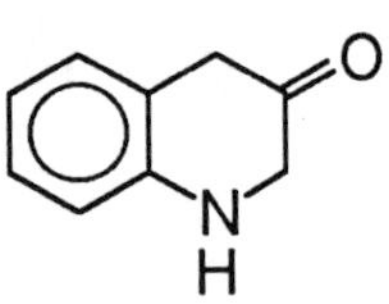

b) 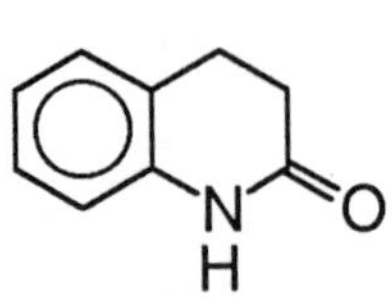

c) 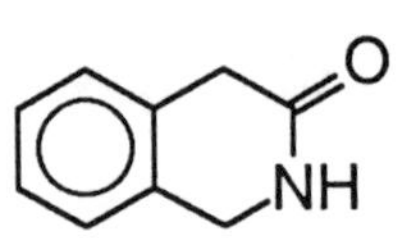

d) 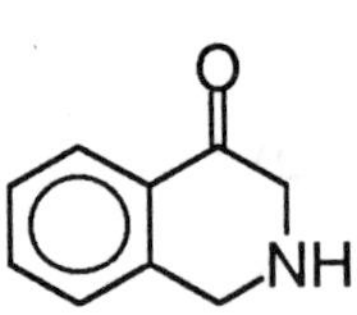

8. What are the correct positions for the most likely site for protonation of the bases, guanidine and imidizole?

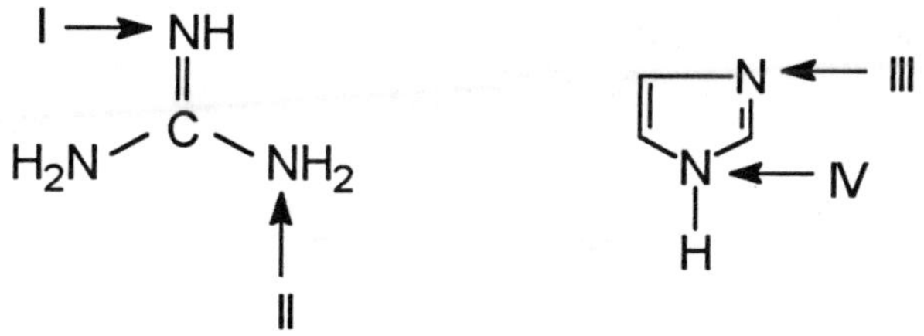

a) I, III b) II, IV c) I, IV d) II, III

9. Which of the following synthetic methods will not work for preparing 2-aminopentane?

a) The Gabriel synthesis using the potassium salt of phthalimide and an alkyl halide.
b) Amination of an alkyl halide in the presence of a large excess of ammonia.
c) Reduction of a secondary amide using $LiAlH_4$.
d) Reduction of an azide using hydrogen and a transition metal catalyst.

10. Which of the following synthetic methods will not work for preparing 1-aminopentane?

a) The Gabriel synthesis using the potassium salt of phthalimide and an alkyl halide.
b) Amination of an alkyl halide in the presence of a large excess of ammonia.
c) Reduction of an amide using $LiAlH_4$.
d) Reductive amination of a ketone and a primary amine by hydrogen and a transition metal catalyst.

11. Which of the following synthetic methods is best for preparing 2-amino-2-methylpropane?

a) The Gabriel synthesis using the potassium salt of phthalimide and an alkyl halide.
b) Amination of a secondary alkyl halide in the presence of a large excess of ammonia.
c) Reduction of a secondary amide using $LiAlH_4$.
d) Reductive amination of a ketone and a primary amine by hydrogen and a transition metal catalyst.

12. Which two reaction sequences prepare the same amine?

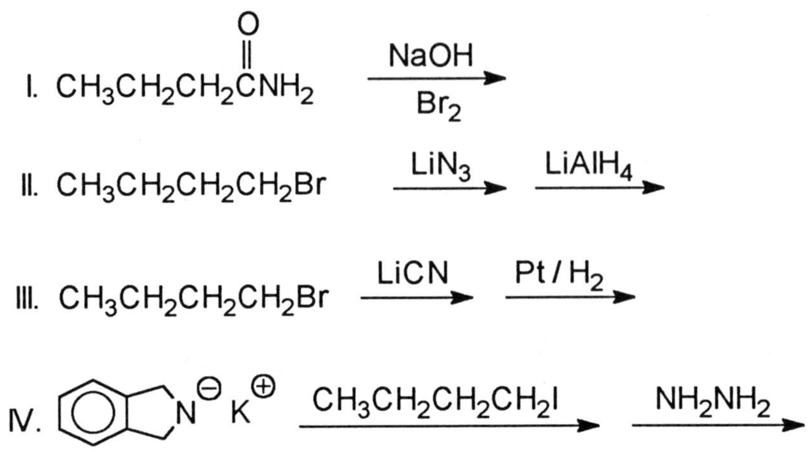

a) I, II b) III, IV c) I, IV d) II, IV

13. What is the major product from the following reaction?

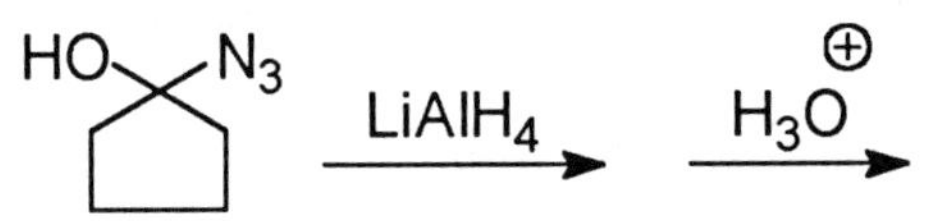

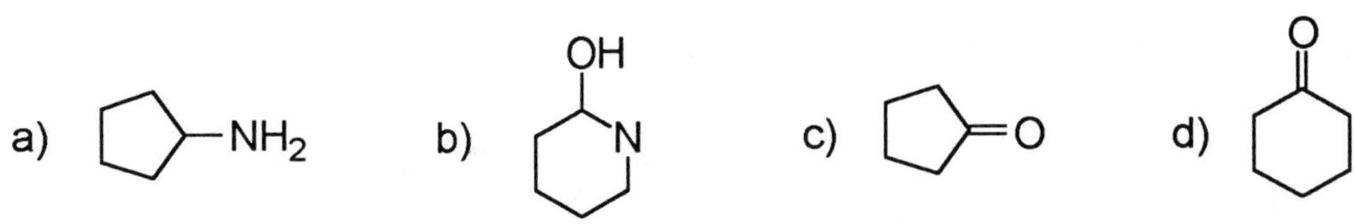

14. Which of the following compounds can be synthesized by using the Sandmeyer reaction?

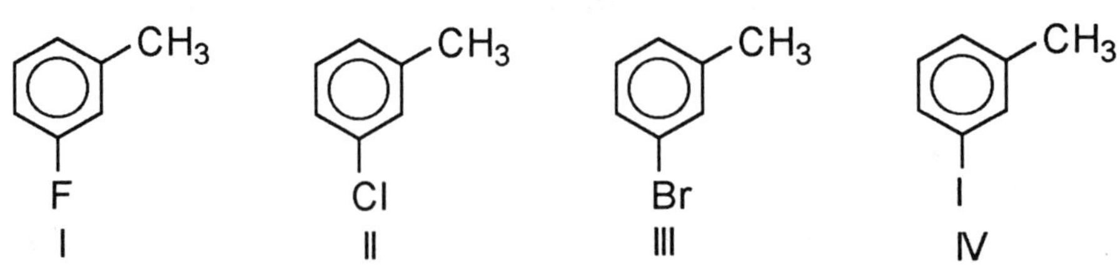

a) I, II b) II, III c) III, IV d) I, IV

15. What are the best reaction conditions for the following transformation?

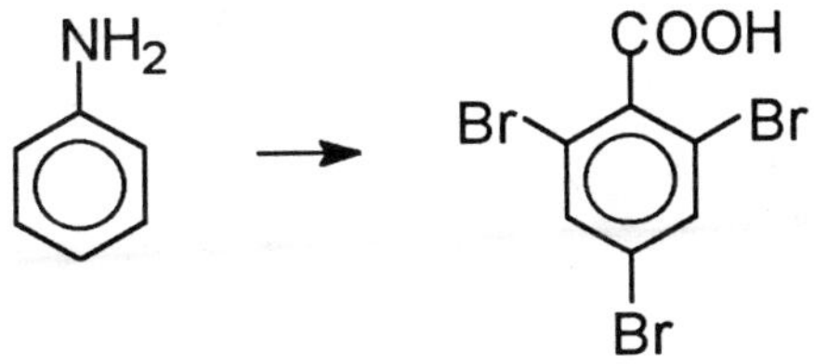

a) $\xrightarrow{Br_2} \xrightarrow[0\text{-}5^{\circ}C]{NaNO_2,\ HCl} \xrightarrow[H_2O]{H_3O^{\oplus}}$

b) $\xrightarrow[0\text{-}5^{\circ}C]{NaNO_2,\ HCl} \xrightarrow{Br_2} \xrightarrow[heat]{CuCN} \xrightarrow[H_2O]{H_3O^{\oplus}}$

c) $\xrightarrow{Br_2} \xrightarrow[0\text{-}5^{\circ}C]{NaNO_2,\ HCl} \xrightarrow[heat]{CuCN} \xrightarrow[H_2O]{H_3O^{\oplus}}$

d) $\xrightarrow[0\text{-}5^{\circ}C]{NaNO_2,\ HCl} \xrightarrow[H_2O]{H_3O^{\oplus}} \xrightarrow{Br_2}$

16. What is the most likely product from the following series of reactions?

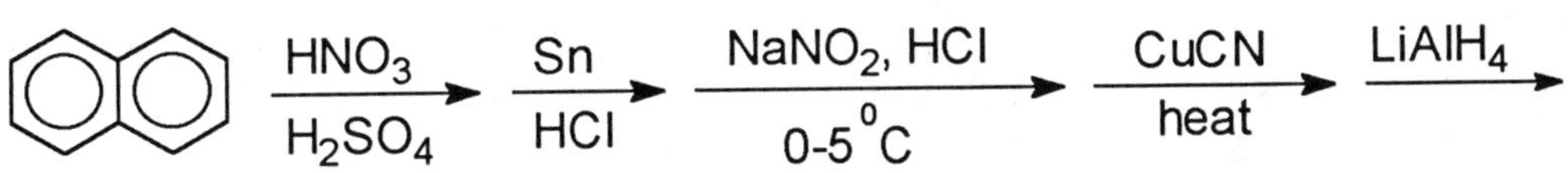

a) 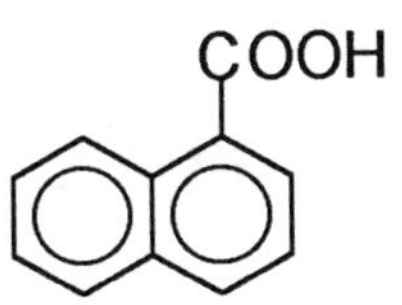

b) 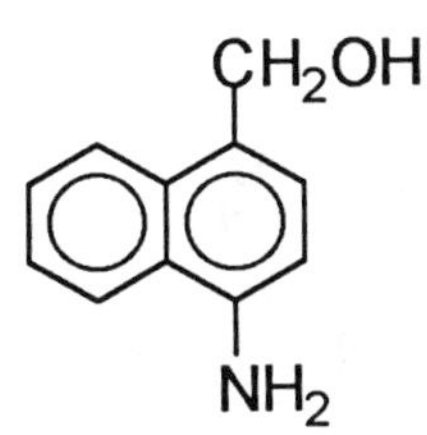

c) 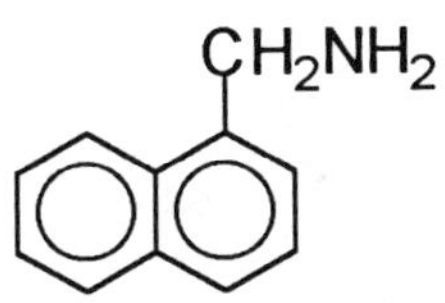

d) 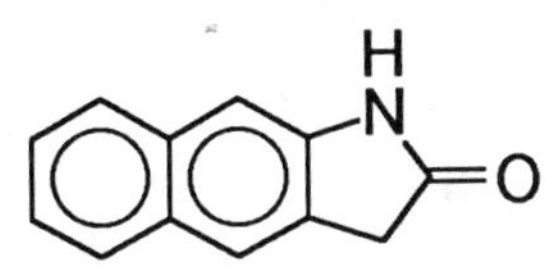

17. What are the best conditions for the following synthesis?

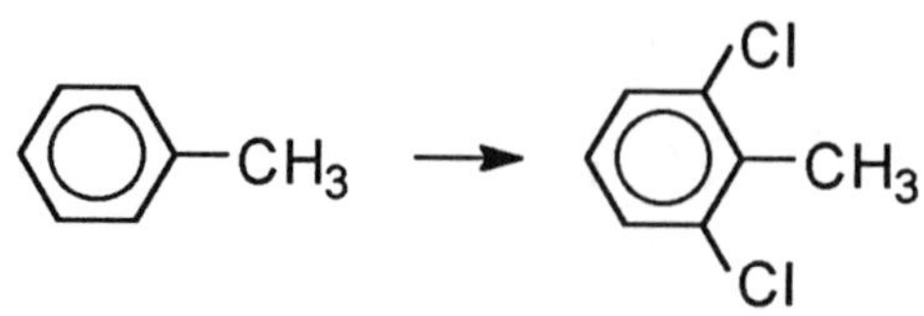

a) $\xrightarrow[H_2SO_4]{HNO_3} \xrightarrow[FeCl_3]{Cl_2} \xrightarrow[HCl]{Sn} \xrightarrow[0\text{-}5^{0}C]{NaNO_2,\ HCl} \xrightarrow{H_3PO_2}$

b) $\xrightarrow[H_2SO_4]{HNO_3} \xrightarrow[HCl]{Sn} \xrightarrow[FeCl_3]{Cl_2} \xrightarrow[0\text{-}5^{0}C]{NaNO_2,\ HCl} \xrightarrow{H_3PO_2}$

c) $\xrightarrow[FeCl_3]{Cl_2} \xrightarrow[H_2SO_4]{HNO_3} \xrightarrow[HCl]{Sn} \xrightarrow[0\text{-}5^{0}C]{NaNO_2,\ HCl} \xrightarrow{H_3PO_2}$

d) $\xrightarrow[H_2SO_4]{HNO_3} \xrightarrow[HCl]{Sn} \xrightarrow[0\text{-}5^{0}C]{NaNO_2,\ HCl} \xrightarrow{H_3PO_2} \xrightarrow[FeCl_3]{Cl_2}$

18. In the Schlieman reaction, a diazonium compound is converted to an aromatic fluorine compound by using which reagent?

a) KF   b) CuF   c) HBF_4 d) NaF

19. Compound A has the molecular formula $C_5H_{13}N$. It fails to react with acetic anhydride or with benzene sulfonyl chloride. Its proton NMR spectrum consists of 3 peaks: a six-proton singlet, a six-proton doublet, and a one-proton multiplet. What is the most likely structure for compound A?

a) $(CH_3)_2NCH_2CH_2CH_3$ b) $(CH_3CH_2)_2NCH_3$

c) $(CH_3)_2NCH(CH_3)_2$ d) $CH_3NHCH_2CH(CH_3)$

20. What is the principal product from the following series of reactions?

$CH_3CH_2CH(NH_2)CH_3 \xrightarrow[CH_3I]{\text{excess}} \xrightarrow[H_2O]{Ag_2O} \xrightarrow{\text{heat}}$

a) $CH_3CH_2CH{=}CH_2$ b) $CH_3C(H){=}C(H)CH_3$ (H's on opposite sides)   c) $CH_3C(H){=}C(H)CH_3$ (H's on same side)   d) $(CH_3)_2C{=}CH_2$

21. An unknown compound has the molecular formula $C_9H_{17}N$. It reacts with excess methyl iodide to give another compound, $C_{10}H_{20}NI$. Treatment of this compound with silver oxide, followed by heating, converts it into another compound, $C_{10}H_{19}N$. When this last compound is treated successively with methyl iodide and silver oxide and then heated, two compounds are generated, C_8H_{12} and C_3H_9N. What is a likely structure for the original unknown compound?

a) 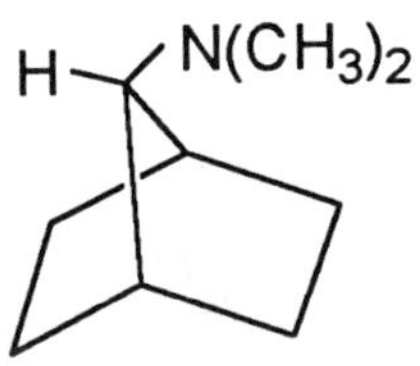

b) 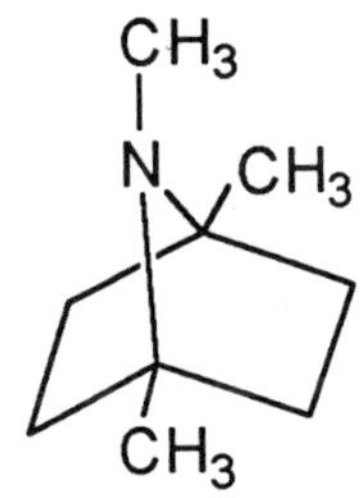

c) 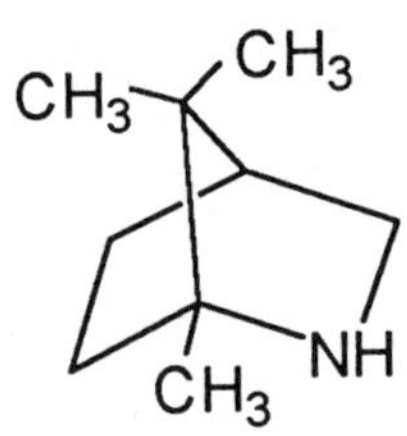

d) 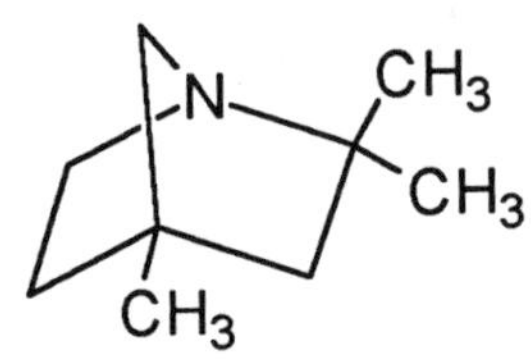

22. The following reaction sequence represents a double exhaustive methylation-Hoffmann elimination scheme. What are the most likely structures for the final product?

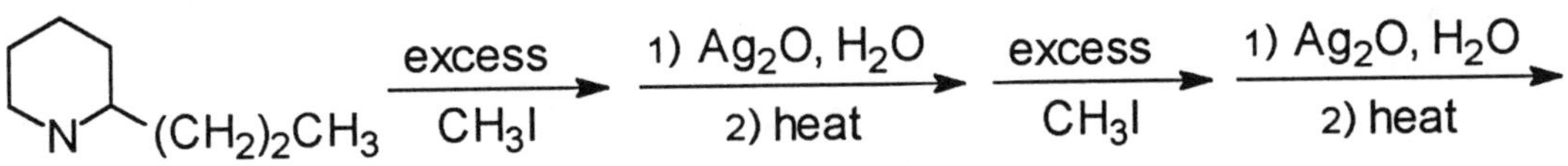

I. 1,4-octatriene II. 1,5-octadiene III. 1,6-octadiene IV. 1,7-octadiene

a) I, II b) II, III c) III, IV d) II, IV

23. Which of the following amines will react with hydrogen peroxide and then undergo a Cope elimination?

a) $(CH_3)_3CCH_2N(CH_3)_2$

b) $CH_3CH_2NHCH_3$

c) $(CH_3)_3N$

d) $(CH_3)_3CN(CH_3)_2$

24. What is the major product from the following reaction?

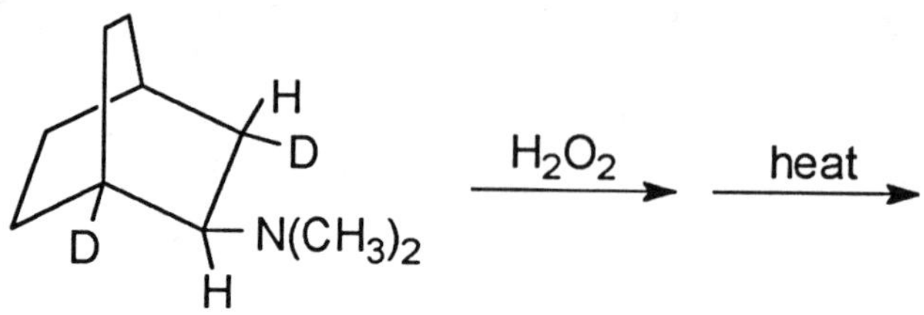

a) 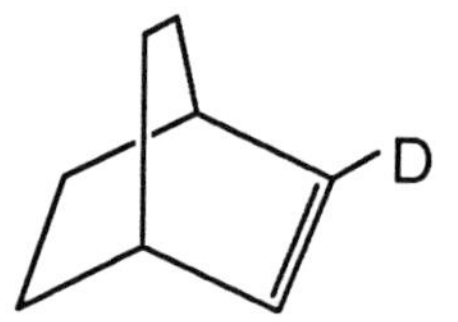

b) 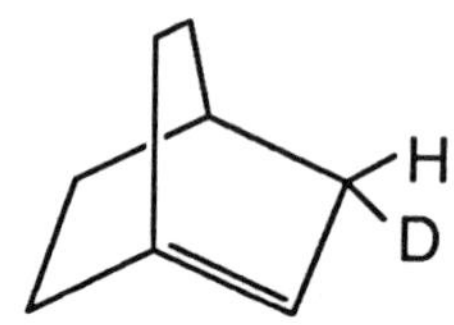

c) 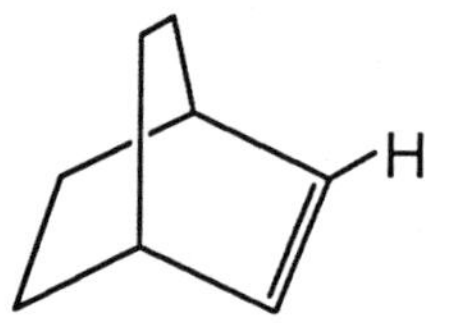

d) 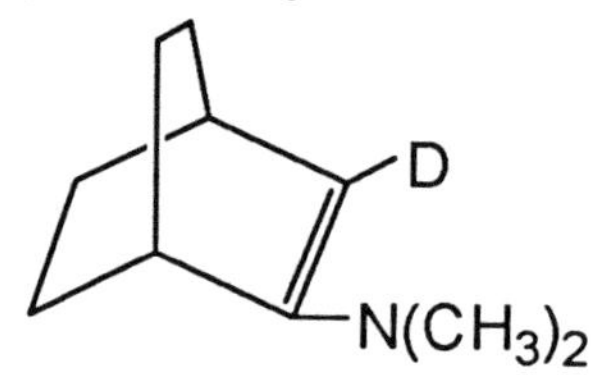

25. An unknown amine is treated with benzene sulfonyl chloride in excess aqueous sodium hydroxide and no solid precipitates until the solution is acidified. When the amine is treated with a solution of 2-methoxy-4-nitrobenzenediazonium ion, a bright orange solid precipitates. What is the most likely structure for the amine?

a) 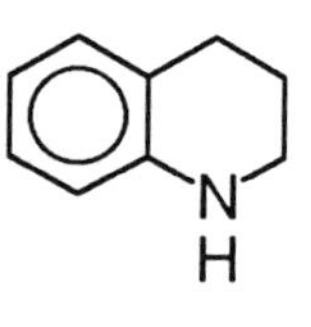

b) 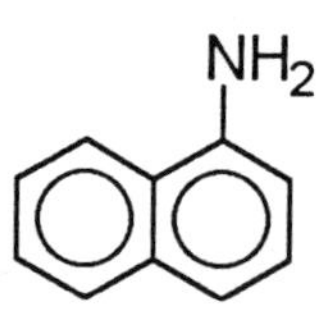

c) 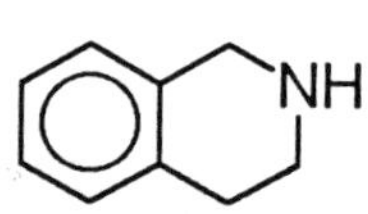

d) 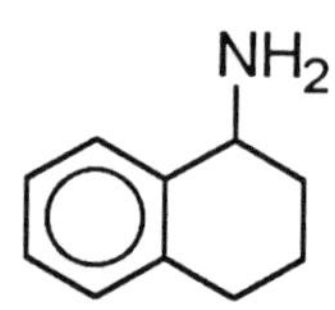

Answers

1. b)
2. d)
3. d)
4. a)
5. c)
6. c)
7. d)
8. a)
9. c)
10. d)
11. d)
12. d)
13. c)
14. b)
15. c)
16. c)
17. a)
18. c)
19. c)
20. a)
21. b)
22. a)
23. d)
24. a)
25. b)

Chapter 23 Lipids

1. Which of the following saturated fatty acids have the correct name?

$CH_3(CH_2)_{10}COOH$ stearic acid I	$CH_3(CH_2)_{14}COOH$ palmitic acid II
$CH_3(CH_2)_{16}COOH$ lauric acid III	$CH_3(CH_2)_{18}COOH$ arachidic acid IV

a) I, II　　b) III, IV　　c) I, III　　d) II, IV

2. What properties are characteristic for the most abundant fatty acids found in plants and animals?

I. They contain an even number of carbon atoms, in the range of 10-20.
II. The Z isomer predominates.
III. The unsaturated fatty acids have a higher melting point than the corresponding saturated acids.
IV. The most abundant fatty acids are lauric, myristic and linolenic.

a) I, II　　b) III, IV　　c) I, III　　d) II, IV

3. What physical property of triacylglycerols rich in unsaturated fatty acids is responsible for the fact that they are usually liquids at room temperature?

a) Van der Waals attractions　　b) hydrophobic interactions
c) cis unsaturated carbon-carbon bonds　　d) hydrogen bonding

4. Washing polyester materials in detergents is more desirable than in soap because of which property?

a) Soaps are more acidic.
b) Soaps are more basic.
c) Detergents have soluble calcium salts.
d) Soap micelles are unstable in acids.

5. The 'ring around the bathtub' is a problem with soaps but not with detergents because of which property?

a) Soaps are more acidic.
b) Soaps are more basic.
c) Detergents have soluble calcium salts.
d) Soap micelles are unstable in acids.

6. How many triglycerides, including stereoisomers, are possible that have two different acyl groups?

a) 2 b) 3 c) 4 d) 5

7. How many triglycerides, including stereoisomers, are possible that have three different acyl groups?

a) 4 b) 6 c) 8 d) 12

8. Which reagent is best for separating estrone and testerone by extraction?

a) aqueous HCl
b) aqueous $NaHCO_3$
c) aqueous KOH
d) Br_2 in CCl_4

9. Which of the following are derivatives of the parent steroid, pregnane?

a) estrone and progesterone
b) progesterone and cholesterol
c) cortisone and progesterone
d) cholic acid and cortisone

10. Which of the following steroids is most acidic?

a) testerone
b) estradiol
c) progesterone
d) cortisone

11. Which functional group accounts for the ability to separate estradiol from progesterone?

a) carboxylic acid b) phenolic OH c) amide d) alcoholic OH

12. What natural products have structures that could be derived from cholesterol?

I. vitamin A II. vitamin D III. cholic acid IV. cortisone V. squalene

a) I, II b) III, IV c) IV, V d) III, V

13. Which of the following groups is NOT characteristic of phospholipids?

a) phosphate esters
b) fatty acid esters
c) glycerides
d) polyamides

14. Which property of phospholipids accounts for their ability to form micelles?

a) nonpolarity
b) unsaturation
c) hydrophilicity and lipophilicity
d) lipophilicity

15. Which property of phospholipids accounts for their ability to form fluid membranes?

a) nonpolarity
b) unsaturation
c) hydrophilicity and lipophilicity
d) lipophilicity

16. The following are key steps in fatty acid metabolism. What is the best description for the first step (A)?

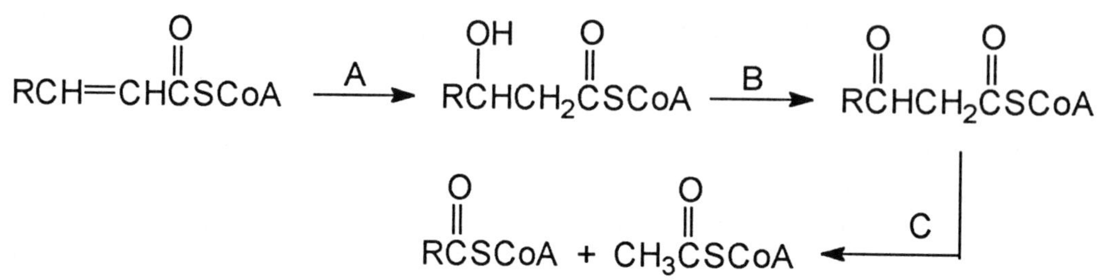

a) oxidation
b) Michael addition
c) retro-Claisen reaction
d) selective reduction

17. The following are key steps in fatty acid metabolism. What is the best description for the second step (B)?

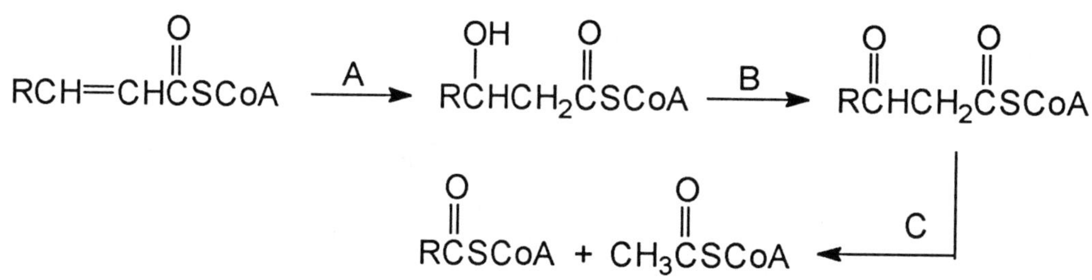

a) oxidation
b) Michael addition
c) retro-Claisen reaction
d) selective reduction

18. The following are key steps in fatty acid metabolism. What is the best description for the third step (C)?

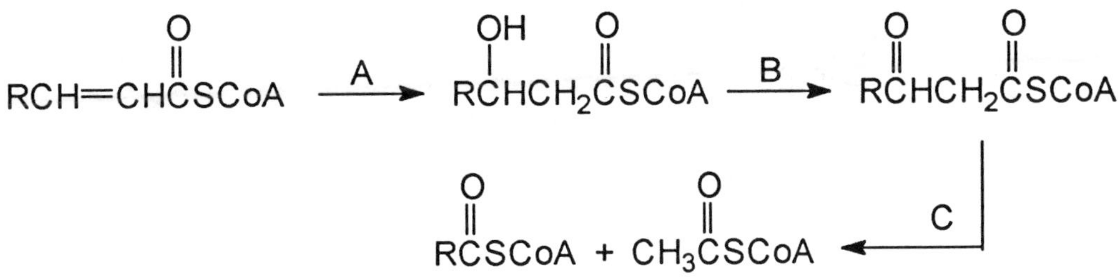

a) oxidation
b) Michael addition
c) retro-Claisen reaction
d) selective reduction

19. The biosynthesis of butyric acid proceeds via a Claisen-like reaction of acetylSCoA. If the starting material is labeled at the methyl carbon atom, which positions will have the label in the product?

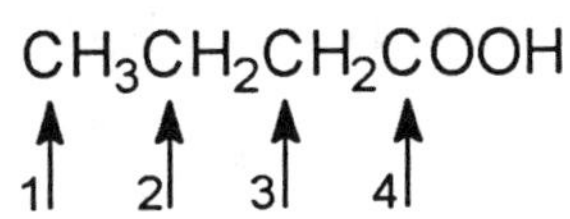

a) 1, 2
b) 1, 3
c) 2, 3
d) 2, 4

20. What is the saponification number of butter fat if the average molecular weight is 730?

a) 77
b) 195
c) 230
d) 690

21. What common structural features of vitamin E and vitamin K_1 account for the greater solubility of these molecules in dichloromethane than in water?

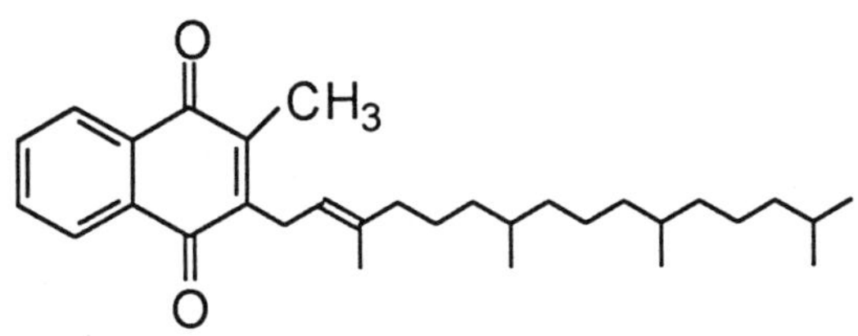

vitamin K_1

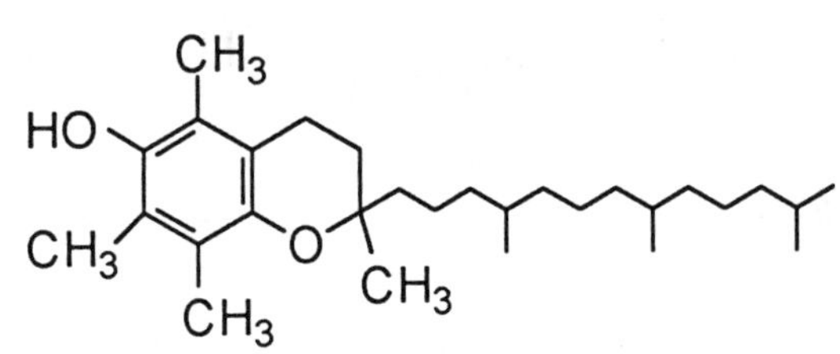

vitamin E

a) aromatic rings
b) oxygen atoms
c) 4 isoprene units
d) a quinone / hydroquinone unit

22. Which of the following structures is a prostaglandin?

a) 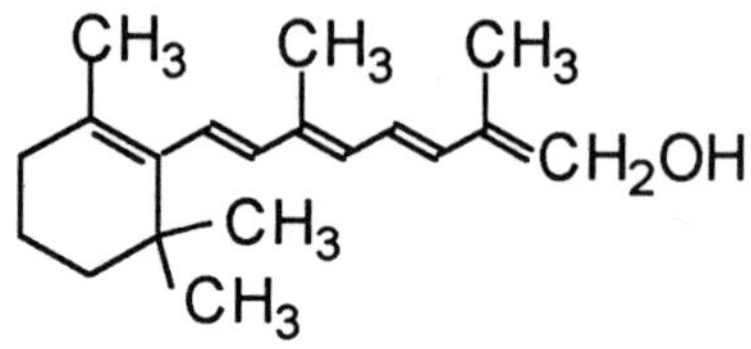

b) 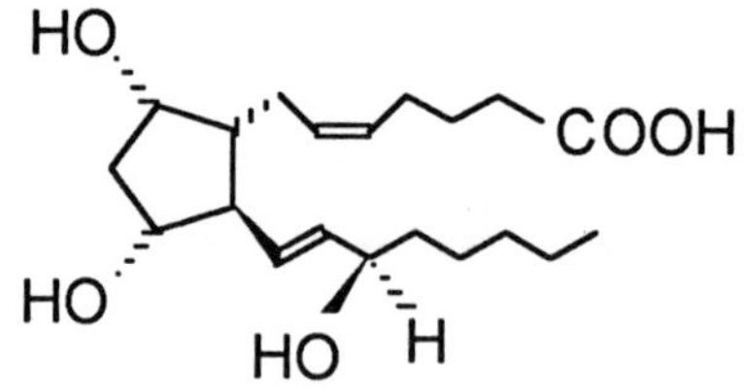

c) 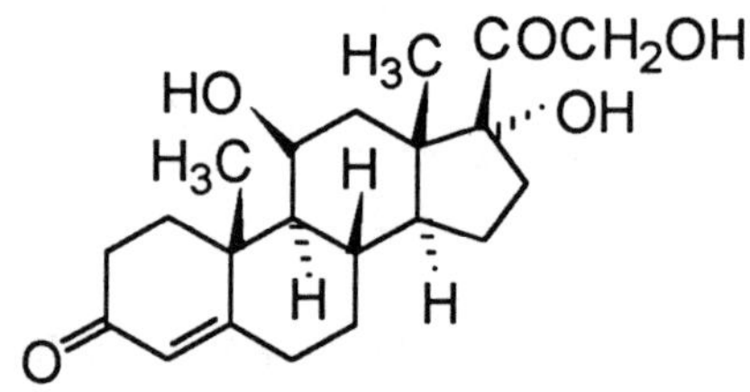

d) 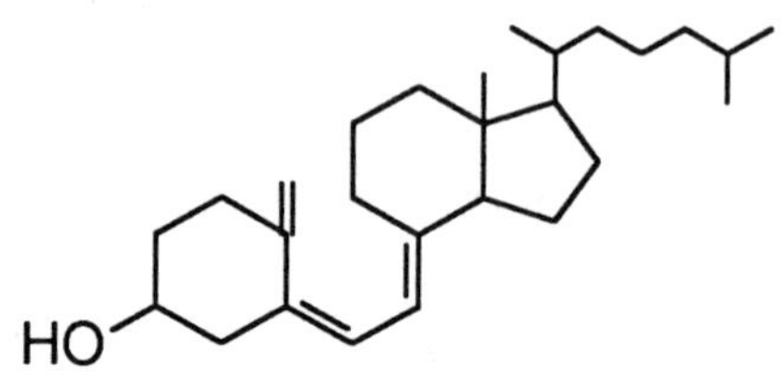

23. Which of the following structures is vitamin A?

a) 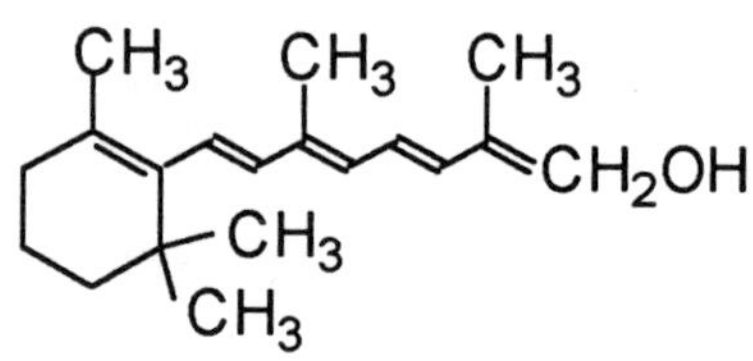

b) 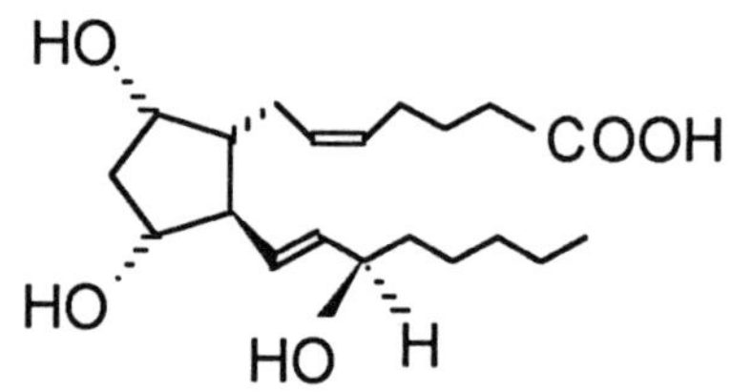

c) 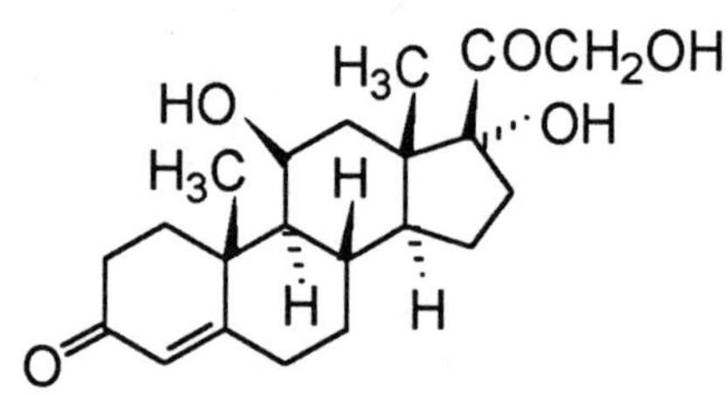

d) 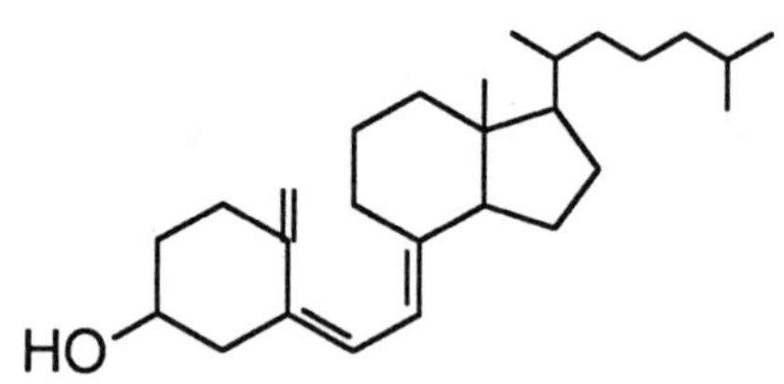

24. Which of the following structures is a cortisone?

a) 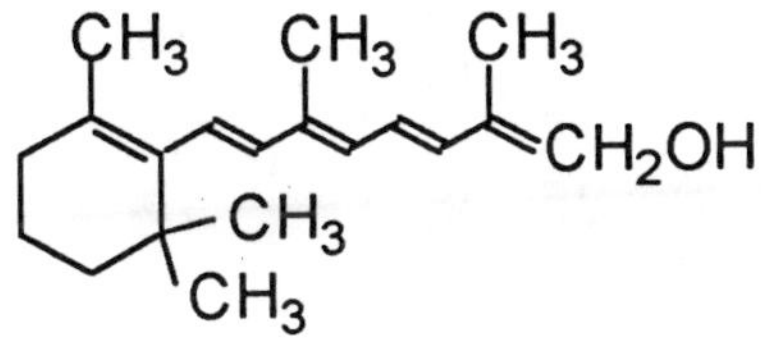

b) 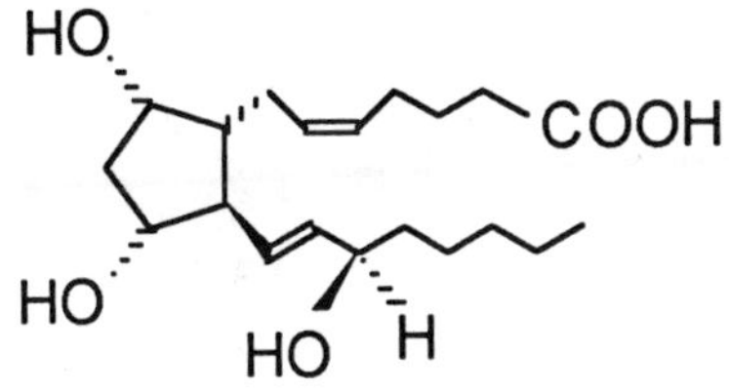

c) 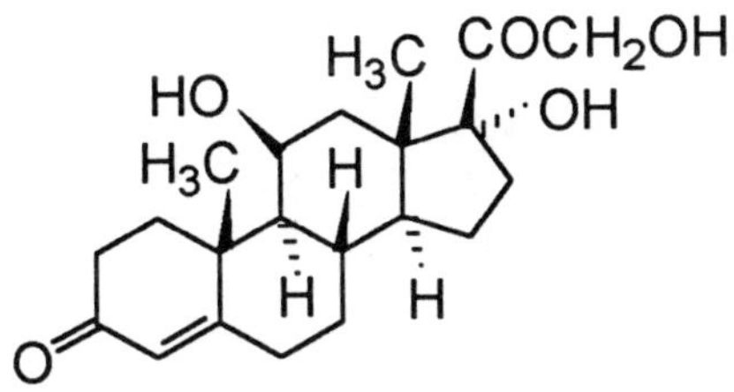

d) 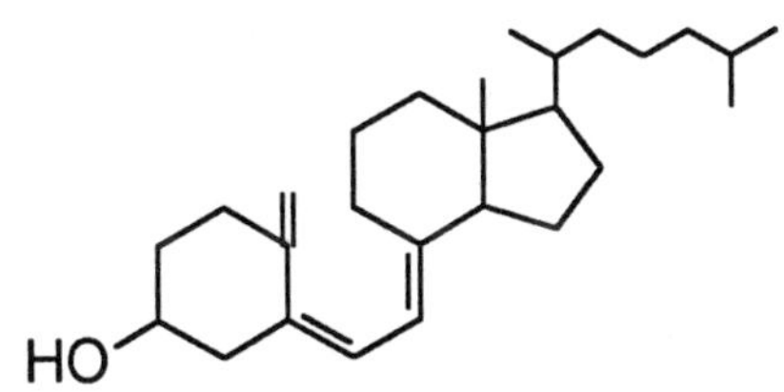

25. Fats, oils, waxes, phospholipids, prostaglandins and steroids all have in common which of the following properties?

I. oxygen functionality
II. nonpolar groups
III. polar groups
IV. unsaturation

a) I, II
b) III, IV
c) I, III
d) II, IV

Answers

1. d)
2. a)
3. c)
4. b)
5. c)
6. b)
7. b)
8. c)
9. c)
10. b)
11. b)
12. b)
13. d)
14. c)
15. c)
16. b)
17. a)
18. c)
19. b)
20. c)
21. c)
22. b)
23. a)
24. c)
25. a)

Chapter 24 Amino Acids and Proteins

1. Which of the following amino acids has a nonpolar side chain?

a) 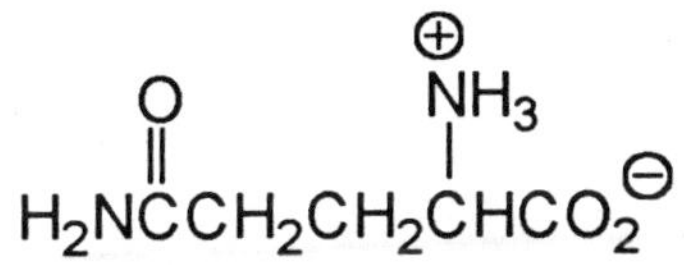

b) 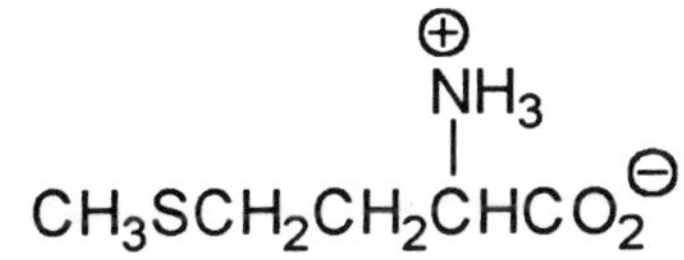

c) 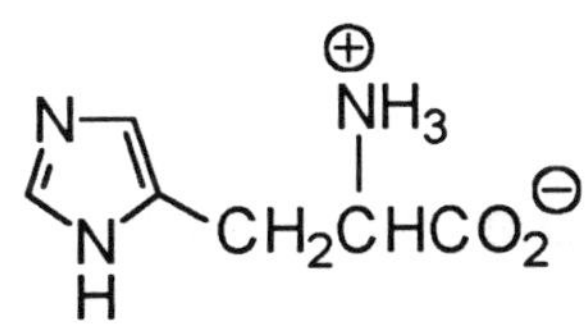

d) 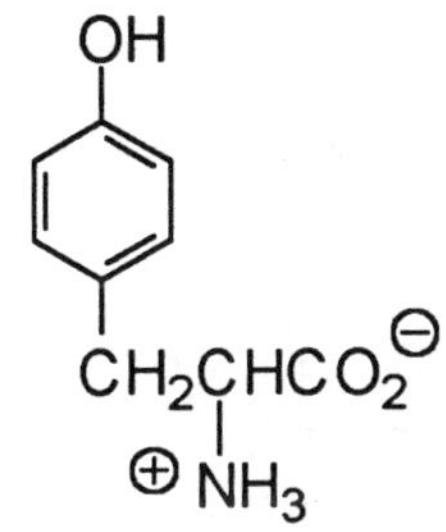

2. Which of the following amino acids has an acidic side chain?

a) 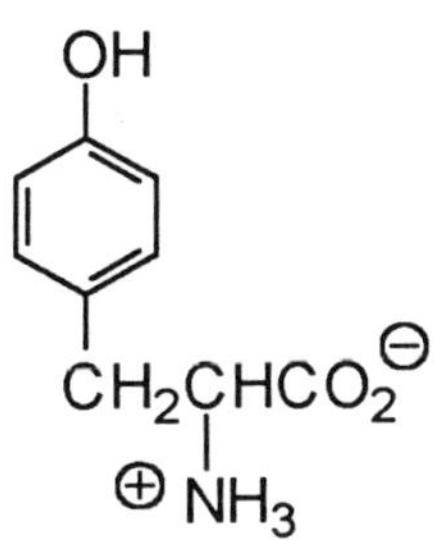

b) 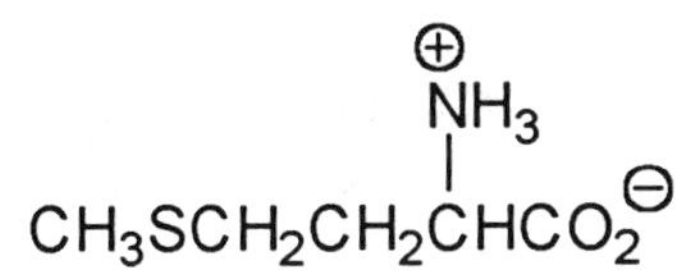

c) 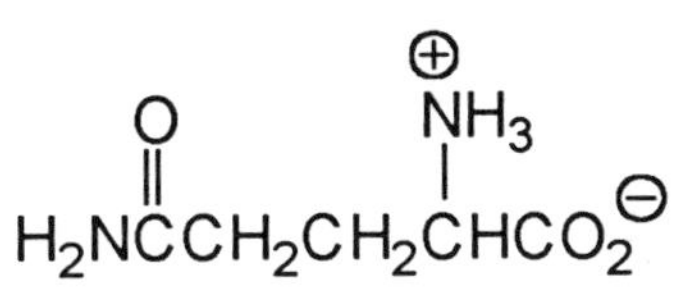

d) 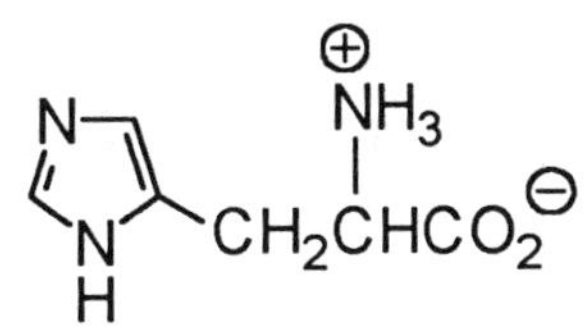

3. Which of the following amino acids has a basic side chain?

a) 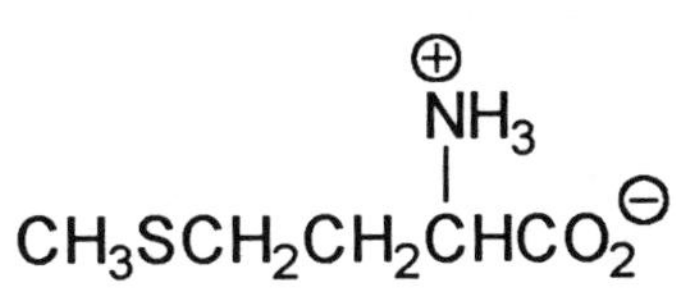

b) 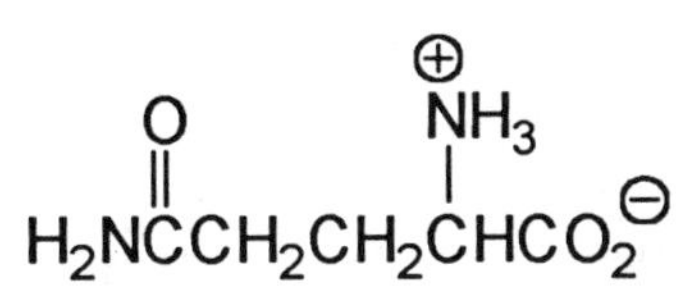

c) 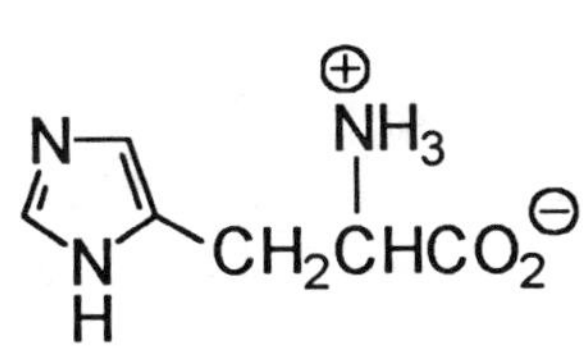

d) 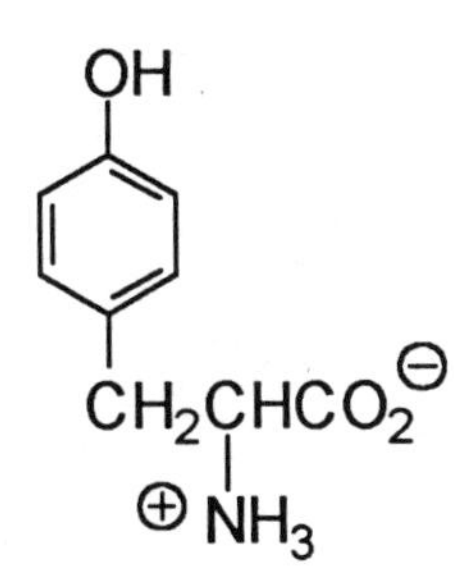

4. How many stereoisomers are possible for the following amino acid?

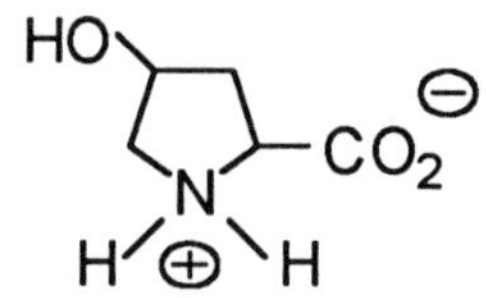

a) 2 b) 4 c) 8 d) none

5. What is the isoelectric point for an amino acid?

a) The degree of ionization of the amino acid at pH 7.
b) The pH at which there is no net charge for the amino acid.
c) The pH at which the ionized form predominates.
d) The pH at which the amino acid is present as the dipolar ion.

6. What is the order of increasing acidity for the indicated carboxyl groups (least first)?

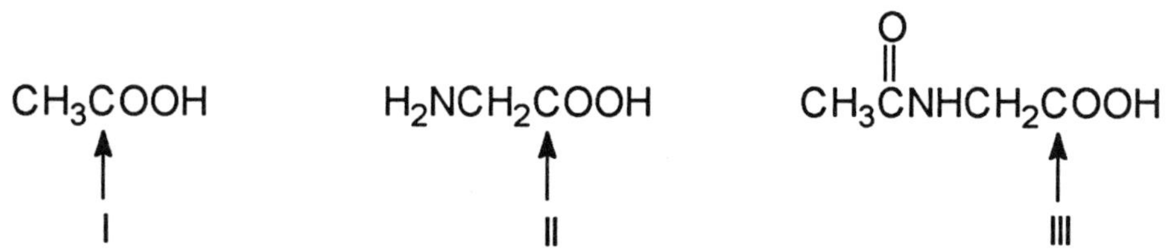

a) I, II, III b) II, III, I c) III, II, I d) I, III, II

7. What is the order of increasing basicity for the indicated amino groups (least first)?

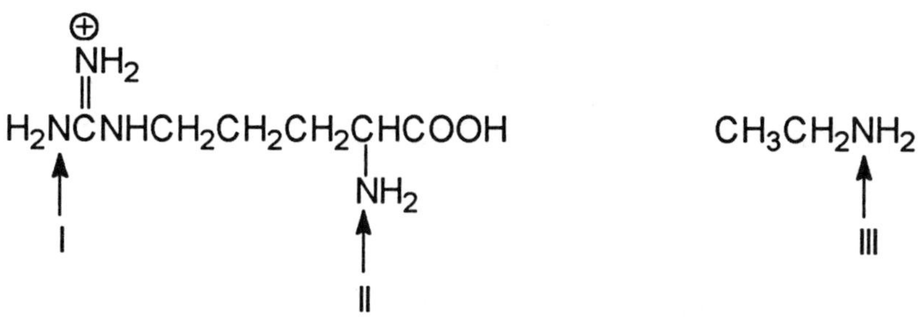

a) I, II, III b) II, III, I c) III, II, I d) I, III, II

8. Tyrosine has an isoelectric point of 5.63. What charge does tyrosine have at pH 7 and which electrode does it migrate to during electrophoresis?

a) negative, anode
b) positive, cathode
c) negative, cathode
d) positive, anode

9. At which pH is the cationic form of glutamic acid present in greatest concentration?

a) 2 b) 3.2 c) 5.6 d) 9.7

10. Which of the following is the predominant form in blood plasma at pH 7.4?

a) $H_3\overset{\oplus}{N}CH_2CH_2CH_2CH_2\underset{}{C}H(\overset{\oplus}{N}H_3)COOH$

b) $H_2NCH_2CH_2CH_2CH_2CH(NH_2)COO^{\ominus}$

c) $H_3\overset{\oplus}{N}CH_2CH_2CH_2CH_2CH(NH_2)COO^{\ominus}$

d) $H_3\overset{\oplus}{N}CH_2CH_2CH_2CH_2CH(\overset{\oplus}{N}H_3)COO^{\ominus}$

11. What is the charge on lysine at pH 11?

a) 0 b) +1 c) -1 d) +2

12. What is the charge on lysine at pH 2?

a) 0 b) +1 c) -1 d) +2

13. What is the structure of the pentapeptide that gave lys-leu-phe on reaction with cyanogen bromide, and gave the fragments met-lys, leu-phe and arg on reaction with trypsin.

a) arg-met-phe-leu-lys
b) lys-leu-phe-arg-met
c) arg-met-lys-leu-phe
d) met-arg-lys-leu-phe

14. A polypeptide gave the following proportions of amino acids upon hydrolysis:

ala (1) gly (1) ileu (2) phe (2) ser (1) try (2)

Treatment of the peptide with 2,4-dinitrofluorobenzene, followed by hydrolysis, gave N-(2,4-dinitrophenyl)tryptophan in the mixture of amino acids. Partial hydrolysis of the peptide gave a mixture of smaller peptides including:

ala-gly gly-try ileu-ileu phe-ileu ser-phe try-phe try-ser ileu-ala

Which of the following is the terminal amino acid (with a free carboxyl group) in the original peptide?

a) try b) ser c) phe d) ileu

15. A polypeptide has the following proportions of amino acids:

gly (1) leu (1) val (1) phe (2)

Treatment of the pentapeptide with chymotrypsin gave the following fragments:

gly-val leu phe

What is the structure of the pentapeptide?

a) phe-phe-gly-val-leu
b) gly-val-phe-leu-phe
c) leu-gly-val-phe-phe
d) phe-leu-gly-val-phe

16. What are the most likely products from the following reaction sequence?

$$\text{glycine} \xrightarrow[\text{NaOH}]{\text{BOC—Cl}} \xrightarrow{\text{valine}} \xrightarrow[CH_3COOH]{HBr}$$

a) gly-val and val-val
b) val-gly and val-val
c) gly-gly and gly-val
d) gly-val and val-gly

17. Which of the following are suitable N-protected glycines useful in peptide synthesis?

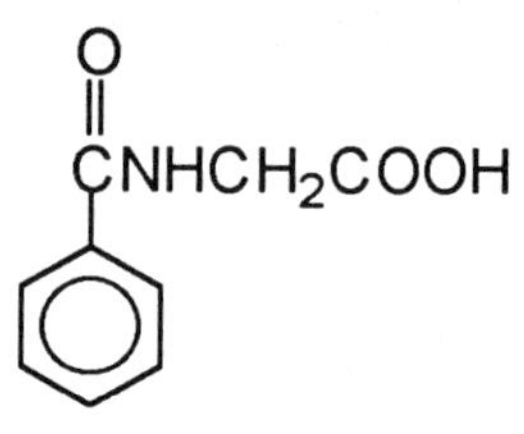

I

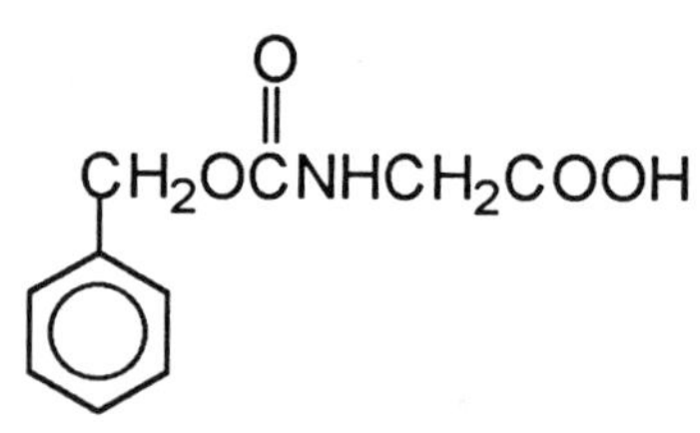

II

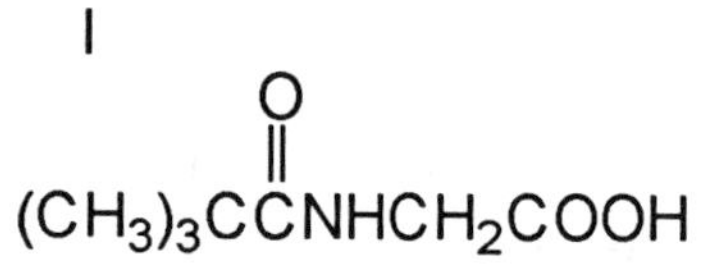

III

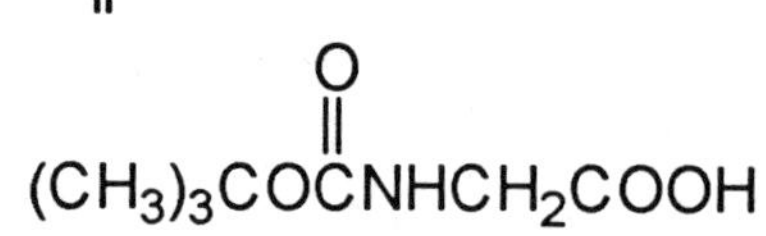

IV

a) I, III
b) II, III
c) I, IV
d) II, IV

18. Which of the following is a useful carboxyl protective group for polypeptide synthesis?

a) benzyl alcohol
b) benzyloxycarbonyl chloride
c) tert- butyloxycarbonyl chloride
d) dicyclohexylcarbodiimide

19. What is the most likely product from the following reaction?

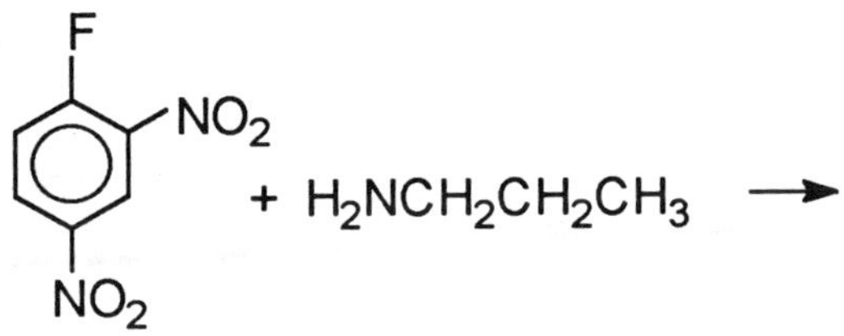

a) 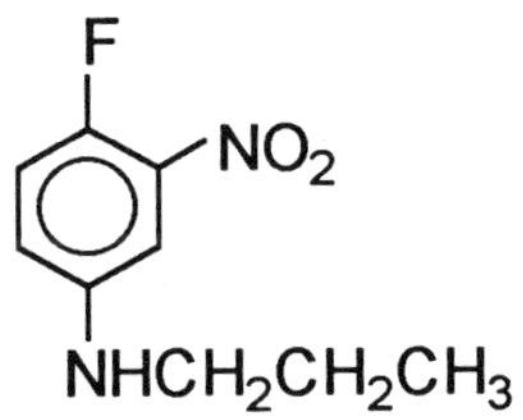

b) 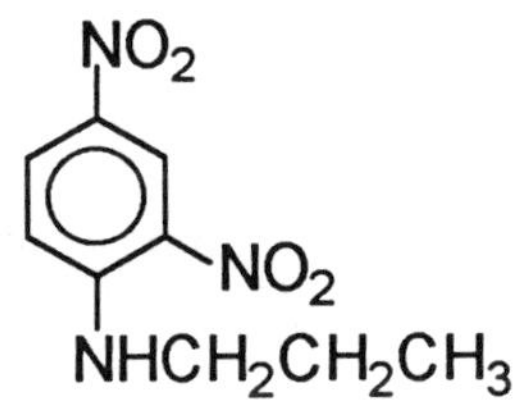

c) 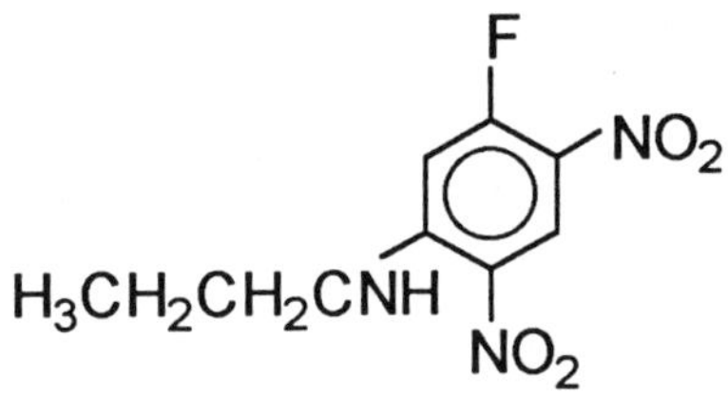

d) 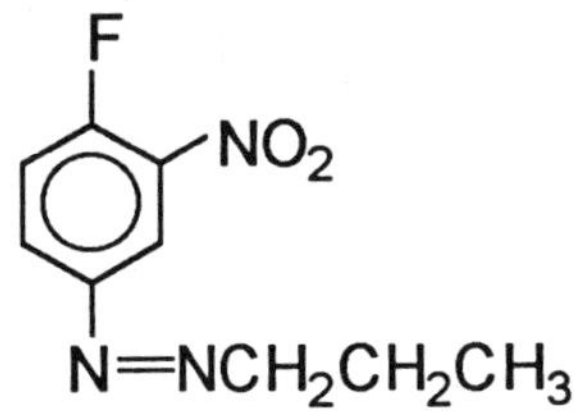

20. What is the final product (C) in the following peptide synthesis?

phenylalanine + MeOH + $H_2SO_4 \longrightarrow$ A

alanine + $C_6H_5CH_2OCOCl$ + NaOH $\longrightarrow$ B

$$A + B \xrightarrow{DCC} C$$

a) Z-ala-phe b) Z-phe-ala-me c) phe-ala d) Z-ala-phe-me

21. In the following structure, how can the carboxyl group be deprotected without deprotecting the amino group?

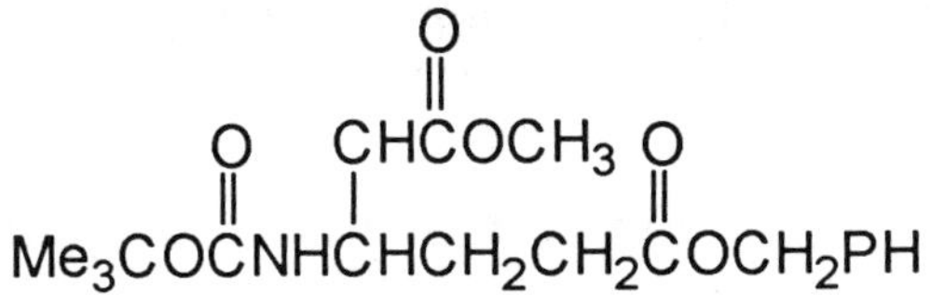

a) dilute HCl b) CF_3COOH c) NaOH d) Pt / H_2

22. In a Merrifield solid phase peptide synthesis, if there are 5 steps and each goes in 90% yield, what is the overall yield ?

a) 18% b) 45% c) 59% d) 90%

23. What is the chemical structure of the tripeptide, glutathione (glu-cys-gly)?

a) 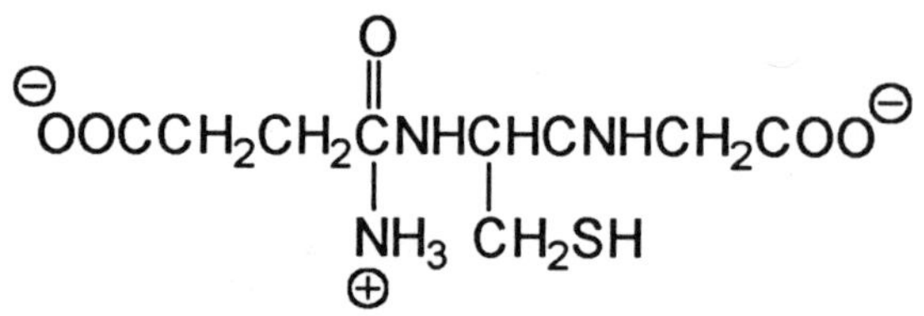

b) 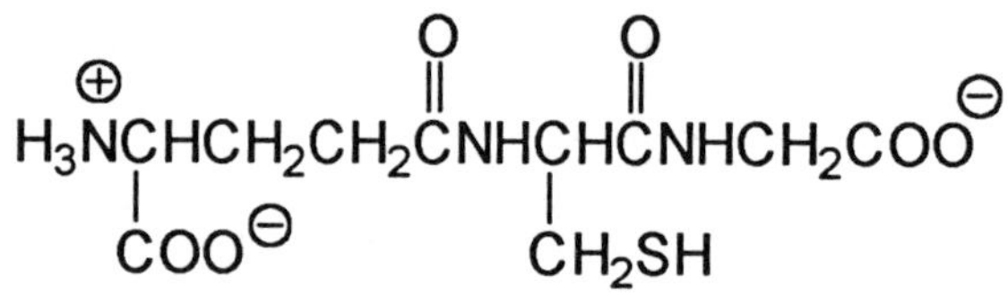

c) 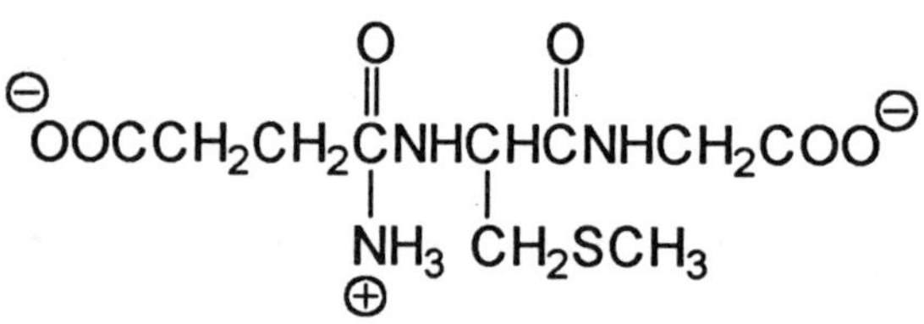

d) 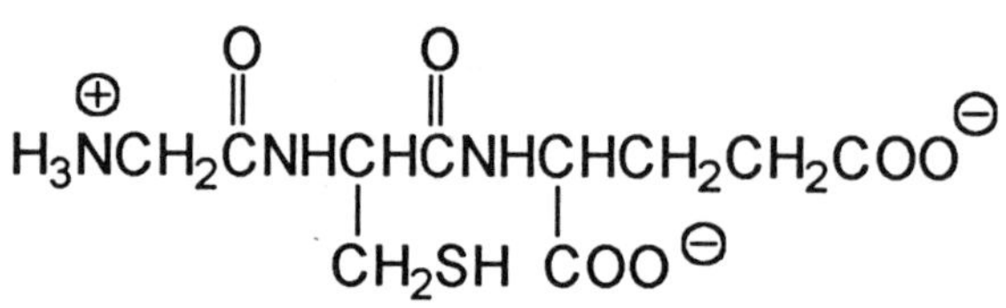

24. The secondary structure of proteins depends primarily on which property of amino acids?

a) disulfide bonds b) hydrogen bonds c) amide bonds d) polar side chains

25. The tertiary structure of proteins depends primarily on which property of amino acids?

a) disulfide bonds b) hydrogen bonds c) amide bonds d) polar side chains

Answers

1. b)
2. a)
3. c)
4. b)
5. b)
6. d)
7. b)
8. a)
9. a)
10. d)
11. c)
12. d)
13. c)
14. c)
15. b)
16. b)
17. d)
18. a)
19. b)
20. b)
21. d)
22. c)
23. b)
24. b)
25. d)

Chapter 25 Nucleic Acids

Which of the following are named correctly?

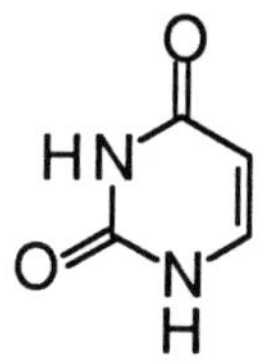

guanine

I

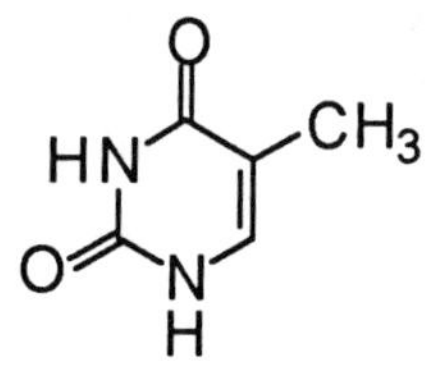

thymine

II

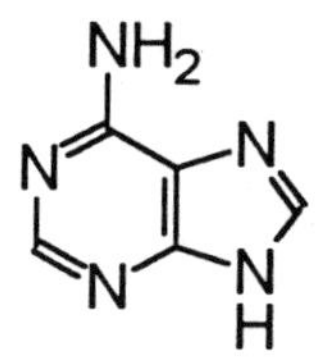

adenine

III

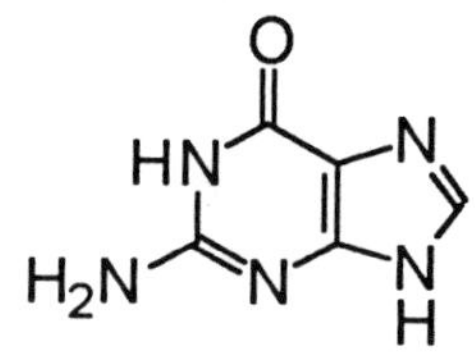

uracil

IV

a) I, II b) II, III c) III, IV d) I, III

2. Which of the following structures is a nucleoside?

a) 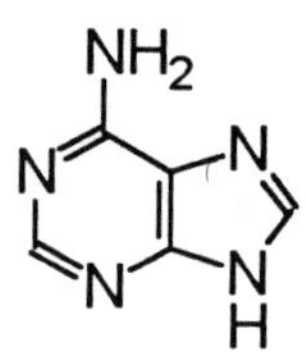

b) 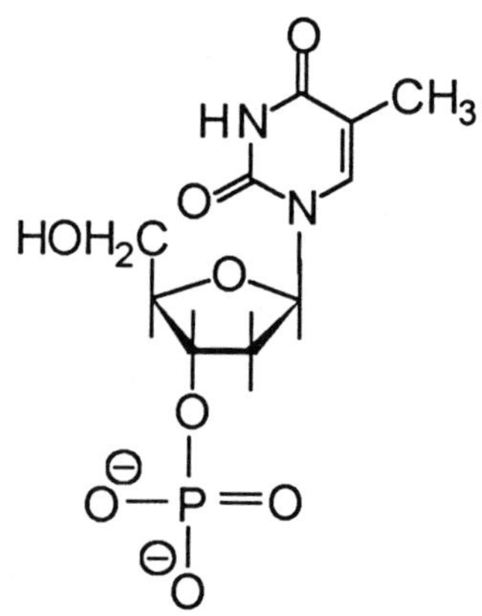

c) 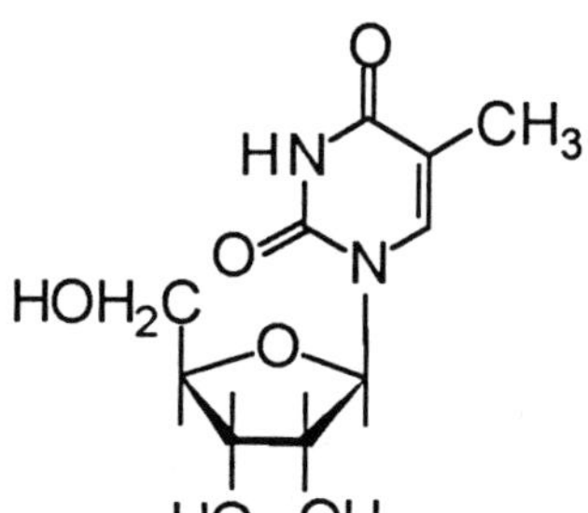

d) 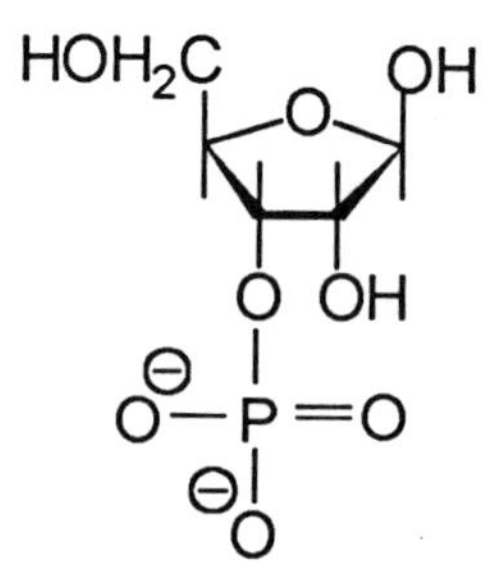

3. What is the correct name for the following structure?

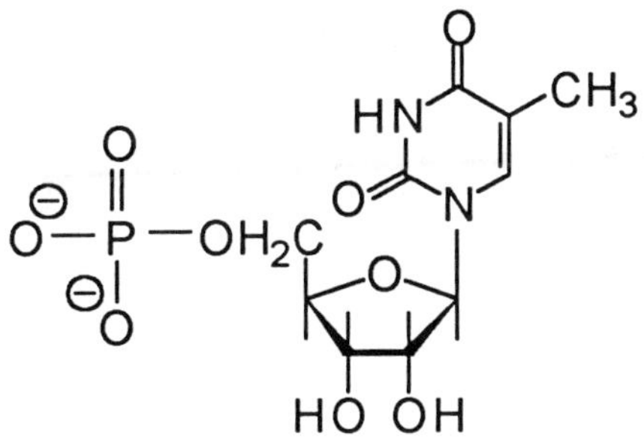

a) 5' dAMP b) 5' TMP c) 3' AMP d) 3' dTMP

4. DNA and RNA differ according to which of the following?

I. the position of attachment of phosphate groups
II. the position of attachment of base groups
III. the sugar structure at C 2'
IV. structure of bases

a) I, II b) II, III c) III, IV d) II, IV

5. What is the correct pairing of bases in the structure for DNA?

A—C	G—T	A—G	G—C	A—T	C—T
I	II	III	IV	V	VI

a) I, II b) III, V c) III, VI d) IV, V

6. In both the RNA and DNA structures, the phosphate groups are attached to the sugar structure at which positions?

a) 5', 5' b) 3', 3' c) 3', 5' d) 2', 5'

7. What is the charge on adenosine phosphate at the blood serum pH of 7.4?

a) 0 b) -1 c) -2 d) -3

8. Which functional groups of DNA do the following structures protect during synthesis?

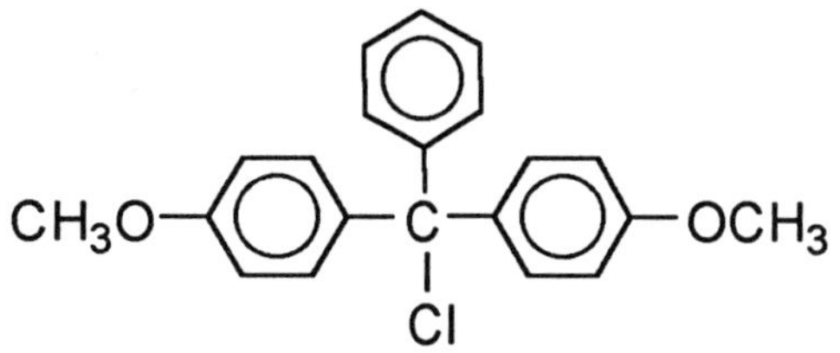

I

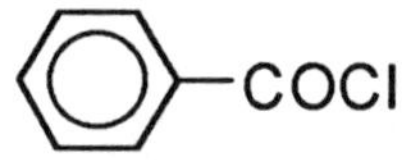

II

a) I for amine, II for 5' hydroxyl
b) II for amine, I for 5' hydroxyl
c) I for 3' nucleoside, II for 5' nucleoside
d) I for 5' nucleoside, II for 3' nucleoside

9. How is removal of the protective groups, shown in the following structure, done in a DNA synthesis?

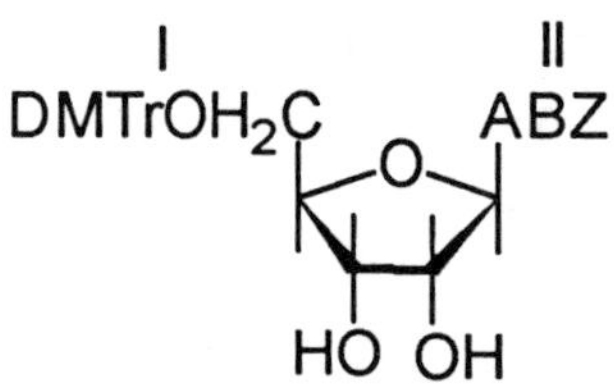

a) NH_3 for I, dichloroacetic acid for II
b) NH_3 for II, dichloroacetic acid for I
c) NaOH for I, HCl for II
d) NaOH for II, HCl for I

10. Nucleosides are rapidly hydrolyzed in which of the following?

a) H_2O b) dil NaOH c) dil HCl d) none of these

11. What is the complementary tetranucleotide for 5'-AGCT-3' ?

a) 3'-TCGA-5'
b) 5'-AGCT-3'
c) 3'-GATC-5'
d) 5'-GATC-3'

12. How many bases are needed to code for an amino acid?

a) 1 b) 2 c) 3 d) 4

13. How is the conversion of uridine to thymidine accomplished?

a) by ring expansion
b) by addition of one carbon atom to the base
c) by deoxygenation at C 2'
d) by cyclization of a phosphodiester

14. A diazotization process can convert which of the following bases?

a) C to U b) G to A c) A to T d) U to G

15. Hydrogen bonding is strongest between which two of the following structures?

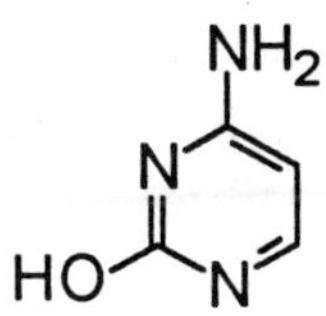
I

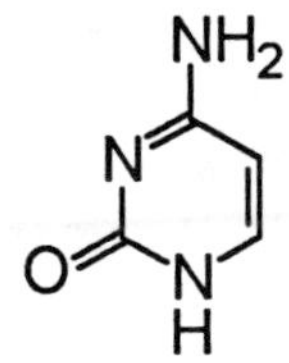
II

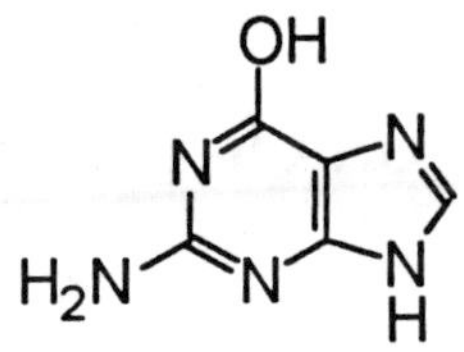
III

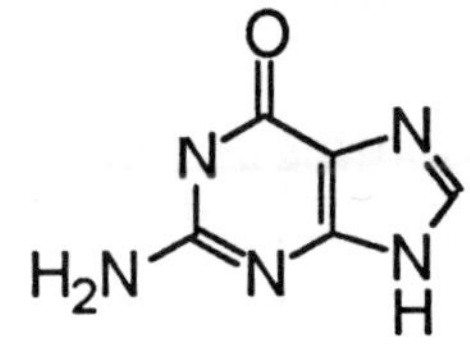
IV

a) I and II b) I and IV c) II and III d) II and IV

16. What is the most likely site of attack of alkylating agents on the DNA structure?

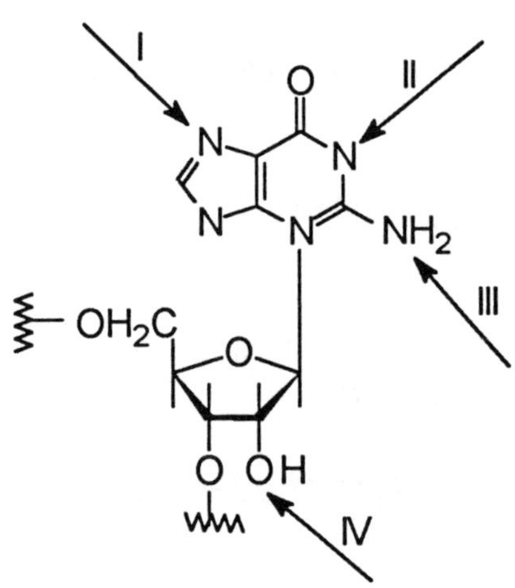

a) I b) II c) III d) IV

17. Hydrolysis of DNA gives which of the following materials?

I. ribose II. adenine III. deoxyribose IV. uracil

a) I, II b) II, III c) III. IV d) I, IV

18. What is the diameter of the double helix form of DNA?

a) 34 Å b) 10 Å c) 20 Å d) 4 m

19. What is the complementary sequence in RNA for the DNA sequence AATCAGTT?

a) AAUCAGUU b) AACUGAUU c) UUAGUCAA d) CCAUCGAA

20. A complete turn of the DNA helix occurs at which of the following distances along the axis?

a) 34 Å b) 10 Å c) 20 Å d) 4 m

21. When RNA is hydrolyzed, which of the following is true for the concentration of the products?

a) A = G = C = U b) (A+G) = (C+U) c) (A+U) = (G+C)
d) There is no relationship.

22. When DNA is hydrolyzed, which of the following is true for the concentration of the products?

a) A = G = C = T b) (A+G) = (C+T) c) (A+T) = (G+C)
d) There is no relationship.

23. Place the following base positions in the order of increasing reactivity in a nucleophilic reaction (least reactive first).

I. N-7 of guanine II. N-3 of adenine III. N-1 of cytosine

a) I, II, III b) II, I, III c) III, I, II d) III, II, I

24. What is the best description for the mechanism of hydrolysis of phosphodiesters?

a) SN_1 b) E_2 c) SN_2 d) E_1

25. The backbone of DNA is best described as:

a) pairs of complementary bases
b) polyphosphate esters of 1,3-glycols
c) pyrophosphates
d) sugar glycosides

Answers

1. b)
2. c)
3. b)
4. c)
5. d)
6. c)
7. c)
8. b)
9. a)
10. c)
11. c)
12. b)
13. b)
14. a)
15. d)
16. a)
17. b)
18. c)
19. c)
20. a)
21. d)
22. c)
23. d)
24. c)
25. b)